KECK FUTURES INITIATIVE

THE FUTURE OF HUMAN HEALTHSPAN

Demography, Evolution, Medicine, and Bioengineering

TASK GROUP SUMMARIES

Conference
Arnold and Mabel Beckman Center
Irvine, California
November 14-16, 2007

THE NATIONAL ACADEMIES PRESS
Washington, D.C.
www.nap.edu

THE NATIONAL ACADEMIES PRESS   500 Fifth Street, N.W.   Washington, DC 20001

NOTICE: The task group summaries in this publication are based on task group discussions during the National Academies Keck *Futures Initiative* Conference *The Future of Human Healthspan: Demography, Evolution, Medicine, and Bioengineering* held at the Arnold and Mabel Beckman Center in Irvine, California, November 14-16, 2007. The discussions in these groups were summarized by the authors and reviewed by the members of each task group. Any opinions, findings, conclusions, or recommendations expressed in this publication are those of the task groups and do not necessarily reflect the view of the organizations or agencies that provided support for this project. For more information on the National Academies Keck *Futures Initiative,* visit www.keckfutures.org.

Funding for the activity that led to this publication was provided by the W.M. Keck Foundation. Based in Los Angeles, the W.M. Keck Foundation was established in 1954 by the late W.M. Keck, founder of the Superior Oil Company. The Foundation's grant making is focused primarily on pioneering efforts in the areas of medical research, science, and engineering. The Foundation also maintains a Southern California Grant Program that provides support in the areas of civic and community services with a special emphasis on children. For more information, visit www.wmkeck.org.

International Standard Book Number-13:   978-0-309-11559-9
International Standard Book Number-10:   0-309-11559-0

Additional copies of this report are available from the National Academies Press, 500 Fifth Street, N.W., Lockbox 285, Washington, DC 20055; (800) 624-6242 or (202) 334-3313 (in the Washington metropolitan area); Internet, http://www.nap.edu.

Copyright 2008 by the National Academy of Sciences. All rights reserved.

Printed in the United States of America

THE NATIONAL ACADEMIES
Advisers to the Nation on Science, Engineering, and Medicine

The **National Academy of Sciences** is a private, nonprofit, self-perpetuating society of distinguished scholars engaged in scientific and engineering research, dedicated to the furtherance of science and technology and to their use for the general welfare. Upon the authority of the charter granted to it by the Congress in 1863, the Academy has a mandate that requires it to advise the federal government on scientific and technical matters. Dr. Ralph J. Cicerone is president of the National Academy of Sciences.

The **National Academy of Engineering** was established in 1964, under the charter of the National Academy of Sciences, as a parallel organization of outstanding engineers. It is autonomous in its administration and in the selection of its members, sharing with the National Academy of Sciences the responsibility for advising the federal government. The National Academy of Engineering also sponsors engineering programs aimed at meeting national needs, encourages education and research, and recognizes the superior achievements of engineers. Dr. Wm. A. Wulf is president of the National Academy of Engineering.

The **Institute of Medicine** was established in 1970 by the National Academy of Sciences to secure the services of eminent members of appropriate professions in the examination of policy matters pertaining to the health of the public. The Institute acts under the responsibility given to the National Academy of Sciences by its congressional charter to be an adviser to the federal government and, upon its own initiative, to identify issues of medical care, research, and education. Dr. Harvey V. Fineberg is president of the Institute of Medicine.

The **National Research Council** was organized by the National Academy of Sciences in 1916 to associate the broad community of science and technology with the Academy's purposes of furthering knowledge and advising the federal government. Functioning in accordance with general policies determined by the Academy, the Council has become the principal operating agency of both the National Academy of Sciences and the National Academy of Engineering in providing services to the government, the public, and the scientific and engineering communities. The Council is administered jointly by both Academies and the Institute of Medicine. Dr. Ralph J. Cicerone and Dr. Wm. A. Wulf are chair and vice chair, respectively, of the National Research Council.

www.national-academies.org

THE NATIONAL ACADEMIES KECK *FUTURES INITIATIVE* HEALTHSPAN STEERING COMMITTEE

JOHN ROWE, M.D. (Chair) (IOM), Professor of Health Policy and Management, Mailman School of Public Health, Columbia University

STEVEN AUSTAD, Professor, Department of Cellular and Structural Biology, The Sam and Ann Barshop Institute for Longevity and Aging Studies, The University of Texas Health Science Center at San Antonio

VERN L. BENGTSON, AARP University Chair in Gerontology and Professor of Gerontology and Sociology, Leonard Davis School of Gerontology, University of Southern California

ELIZABETH BLACKBURN (NAS/IOM), Morris Herzstein Professor of Biology and Physiology, Department of Biochemistry and Biophysics, University of California, San Francisco.

JUDITH CAMPISI, Senior Scientist, Life Sciences Division, Lawrence Berkeley National Laboratory, Professor, Buck Institute for Age Research

LAURA L. CARSTENSEN, Professor and Vice Chair and Director of the Stanford Center on Longevity, Stanford University

RORY A. COOPER, Distinguished Professor, FISA/PVA Chair, Rehabilitation Science and Technology, University of Pittsburgh

CALEB E. FINCH, ARCO-William F. Kieschnick Professor in the Neurobiology of Aging, and Co-Director, Alzheimer Disease Research Center, University of Southern California

RICHARD J. HODES (IOM), Director, National Institute on Aging, National Institutes of Health

RONALD LEE (NAS), Professor of Demography and Jordan Family Professor of Economics, Director, Center for the Economics and Demography of Aging, Department of Demography, University of California, Berkeley

GEORGE M. MARTIN, M.D. (IOM), Professor of Pathology Emeritus, Director Emeritus Alzheimer's Disease Research Center, University of Washington

ROBERT M. NEREM (NAE/IOM), Parker H. Petit Distinguished Chair for Engineering in Medicine, Institute Professor and Director of the Parker H. Petit Institute for Bioengineering and Bioscience, Georgia Institute of Technology

JAMES RIMMER, Professor and Director, National Center on Physical Activity and Disability, Rehabilitation Engineering Research Center on Recreational Technologies and Exercise Physiology for Persons with Disabilities, Department of Disability and Human Development, University of Illinois at Chicago

MICHAEL ROSE, Professor, Department of Ecology and Evolutionary Biology, University of California, Irvine

KHALED J. SALEH, M.D., Associate Professor Orthopedic Surgery and Health Evaluative Sciences Division, Division Head and Fellowship Director, Adult Reconstruction, University of Virginia

SAVIO L-Y WOO (NAE/IOM), Ph.D, D.Sc, Whiteford Professor and Director, Musculoskeletal Research Center, Department of Bioengineering, University of Pittsburgh

Staff

KENNETH R. FULTON, Executive Director
KIMBERLY SUDA-BLAKE, Program Director
MEGAN ATKINSON, Senior Program Specialist
ANNE HEBERGER, Research Associate
RACHEL LESINSKI, Senior Program Specialist

The National Academies Keck *Futures Initiative*

The National Academies Keck *Futures Initiative* was launched in 2003 to stimulate new modes of scientific inquiry and break down the conceptual and institutional barriers to interdisciplinary research. The National Academies and the W.M. Keck Foundation believe that considerable scientific progress will be achieved by providing a counterbalance to the tendency to isolate research within academic fields. The *Futures Initiative* is designed to enable scientists from different disciplines to focus on new questions, upon which they can base entirely new research, and to encourage and reward outstanding communication between scientists as well as between the scientific enterprise and the public.

The *Futures Initiative* includes the following components:

***Futures* Conferences** bring together some of the nation's best and brightest researchers from academic, industrial, and government laboratories to explore and discover interdisciplinary connections in important areas of cutting-edge research. Each year some 100 outstanding researchers are invited to discuss ideas related to a single cross-disciplinary theme. Participants gain not only a wider perspective but also, in many instances, new insights and techniques that might be applied in their own work. Additional pre- or postconference meetings build on each theme to foster further communication of ideas.

Selection of each year's theme is based on assessments of where the intersection of science, engineering, and medical research has the greatest potential to spark discovery. The first conference, in 2003, explored *Signals, Decisions, and Meaning in Biology, Chemistry, Physics, and Engineering*. The 2004 conference focused on *Designing Nanostructures at the Interface between Biomedical and Physical Systems*. The theme of the 2005 conference was *The Genomic Revolution: Implications for Treatment and Control of Infectious Disease*. In 2006 the conference focused on *Smart Prosthetics: Exploring Assistive Devices for the Body and Mind*. In 2007 the conference explored *The Future of Human Healthspan: Demography, Evolution, Medicine, and Bioengineering*, and in 2008 the conference will focus on *Complexity*.

Futures Grants provide seed funding to *Futures* Conference participants, on a competitive basis, to enable them to pursue important new ideas and connections stimulated by the conferences. These grants fill a critical missing link between bold new ideas and major federal funding programs, which do not currently offer seed grants in new areas that are considered risky or exotic. These grants enable researchers to start developing a line of inquiry by supporting the recruitment of students and postdoctoral fellows, the purchase of equipment, and the acquisition of preliminary data, which in turn can position the researchers to compete for larger awards from other public and private sources.

National Academies Communication Awards are designed to recognize, promote, and encourage effective communication of science, engineering, medicine, and interdisciplinary work within and beyond the scientific community. Each year the *Futures Initiative* awards $20,000 prizes to those who have advanced the public's understanding and appreciation of science, engineering, and/or medicine. Beginning in 2008, the awards will be given in four categories: books, newspaper/magazine, online, and TV/radio/film. The winners are honored during the *Futures* Conference.

NAKFI cultivates science writers of the future by inviting graduate students from six science writing programs across the country to attend the conference and develop task group discussion summaries and a conference overview for publication in this book following the conference. Students are selected by the department director or designee, and prepare for the conference by reviewing the webcast tutorials and suggested reading, and selecting a task group in which they would like to participate. Students then work with the task group to which they're assigned to finish the report following the conferences.

FACILITATING INTERDISCIPLINARY RESEARCH STUDY

During the first 18 months of the Keck *Futures Initiative*, the Academies undertook a study on facilitating interdisciplinary research. The study examined the current scope of interdisciplinary efforts and provided recommendations as to how such research could be facilitated by funding organizations and academic institutions. *Facilitating Interdisciplinary Research* (2005) is available from the National Academies Press (www.nap.edu) in print and free PDF versions.

ABOUT THE NATIONAL ACADEMIES

The National Academies comprise the National Academy of Sciences, the National Academy of Engineering, the Institute of Medicine, and the National Research Council, which perform an unparalleled public service by bringing together experts in all areas of science and technology, who serve as volunteers to address critical national issues and offer unbiased advice to the federal government and the public. For more information, visit www.national-academies.org.

ABOUT THE W.M. KECK FOUNDATION

Based in Los Angeles, the W.M. Keck Foundation was established in 1954 by the late W.M. Keck, founder of the Superior Oil Company. The Foundation's grant making is focused primarily on pioneering efforts in the areas of medical research, science, and engineering. The Foundation also maintains a Southern California Grant Program that provides support in the areas of civic and community services with a special emphasis on children. For more information, visit www.wmkeck.org.

The National Academies Keck *Futures Initiative*
100 Academy
Irvine, CA 92617
949-721-2270 (Phone)
949-721-2216 (Fax)
www.keckfutures.org

Preface

At the National Academies Keck *Futures Initiative* Conference *The Future of Human Healthspan: Demography, Evolution, Medicine, and Bioengineering,* participants were divided into 12 interdisciplinary working groups. The groups spent eight hours over two days exploring diverse challenges at the interface between science, engineering, and medicine. The composition of the groups was intentionally diverse, to encourage the generation of new approaches by combining a range of different types of contributions. The groups included researchers from science, engineering, and medicine, as well as representatives from private and public funding agencies, universities, businesses, journals, and the science media. Researchers represented a wide range of experience—from postdoc to those well established in their careers—from a variety of disciplines that included science and engineering, physical medicine and rehabilitation, biology, materials science, biomedical engineering, electrical engineering, chemistry, neuroscience, pharmacology, anatomy, genetics, and physics.

The groups had to address the challenge of communicating and working together from a diversity of expertise and perspectives as they attempted to solve a complicated interdisciplinary problem in a relatively short time. Each group decided on its own structure and approach to tackling the problem. Some groups decided to refine or redefine their problems based on their experience.

Each group presented two brief reports to the whole conference: (1) an interim report on Thursday to debrief on how things were going, along with any special requests (such as an expert in inflammation) and (2) a final briefing on Friday, when each group:

- provided a concise statement of the problem;
- outlined a structure for its solution;
- identified the most important gaps in science and technology and recommended research areas needed to attack the problem; and
- indicated the benefits to society if the problem could be solved.

Each task group included a graduate student in a university science writing program. Based on the group interaction and the final briefings, the students wrote the following summaries, which were reviewed by the group members. These summaries describe the problem and outline the approach taken, including what research needs to be done to understand the fundamental science behind the challenge, the proposed plan for engineering the application, the reasoning that went into it, and the benefits to society of the problem solution.

Eight webcast tutorials were held in September to help bridge the gaps in terminology used by the various disciplines. Participants had the opportunity to ask questions of the webcast speakers during panel sessions, prior to dividing into their task groups.

Contents

Conference Summary 1

TASK GROUP SUMMARIES

Enhancing the Functional Status of the Future Elderly 7

Design New Research Paradigms to Assess Healthspan,
Its Enhancement, and Prolongation in Experimental
Research Animals 17
    Task Group Summary A, 19
    Task Group Summary B, 24

Effects of Exercise on Human Healthspan 29

The Relationship Between Demographic Mortality Rates,
Aging, and Functional Human Healthspan 41

Changes in Social Contexts to Enhance Functional
Status of the Elderly 49

Develop Technological Interventions to Overcome Barriers to
Independence and Community Participation 57
    Task Group Summary A, 62
    Task Group Summary B, 65

Cellular and Molecular Mechanisms of Biological Aging:
The Roles of Nature, Nurture, and Chance in the
Maintenance of Human Healthspan   73
    Task Group Summary A, 78
    Task Group Summary B, 84
    Task Group Summary C, 88

Inflammation's Effects on Aging   93

APPENDIXES

Preconference Webcast Tutorials   103

Agenda   107

Participants   113

To view the preconference webcast tutorials or conference presentations, please visit our website at www.keckfutures.org.

Conference Summary
Megan Chao

For many centuries, discovering the fountain of youth has been just a dream for humans. Aging is an inevitable process in human life, the result of a highly variable biological cycle. As we grow through childhood, we learn fundamental skills to be functional as adults, but as we progress from adulthood to the end of our days, the possibility of going into a functional decline becomes a great risk. It is hard to think about our lives in the long term, and even harder to imagine that maybe one day we could be immobilized in bed, permanently attached to an oxygen tank, or needing assistance from others. The independence we have spent a great deal of our lives finding could be gone in just a short period of time.

The elusive nature of the aging process and finding new ways of addressing human healthspan brought more than 150 experts and researchers from public and private institutions across the nation and around the globe to the Arnold and Mabel Beckman Center in Irvine, California. From November 14 to 16, 2007, it was a convergence of great minds for the fifth annual conference of the National Academies Keck *Futures Initiative* (NAKFI). Attendees from a wide range of fields, including public health, bioengineering, gerontology, and neuroscience, challenged this year's topic, *The Future of Human Healthspan: Demography, Evolution, Medicine, and Bioengineering.*

MEETING IN THE MIDDLE

The prospect of interdisciplinary collaboration drove much of the enthusiasm and excitement of participants at the conference, but a lot of work had to be done before arriving in sunny southern California. Because researchers and experts came from a multitude of disciplines, it was important to bridge any gaps in understanding in order to provide the most productive conference possible. Webcast tutorials were broadcasted online in late September, eight in total, designed to help attendees understand terminology used by researchers in different fields.

Kicking off the first of the webcasts was Dr. Ken Wachter, professor and chair of demography and statistics at the University of California, Berkeley. He presented a tutorial on the demography of aging and the process of extending life expectancy. This was important in painting a picture for an overall understanding of trends in aging, not only in the United States but also around the world. Teresa Seeman, Ph.D., professor of medicine and epidemiology at the University of California, Los Angeles, followed with a presentation on stress and lifestyle and how those factors may influence the decline of health at older ages.

Other webcasts included the integration of technology in the quality of life, gerontology, regenerative medicine, animal models in research, and life expectancy with regard to social and behavioral traits.

CHALLENGING AGING

The chair of this year's conference, Dr. Jack Rowe, professor in the Department of Health Policy and Management at Columbia University, summed up the nature of the conference in his opening remarks.

"I'm sure I'm not the only one who's not sure what's going to happen," he said, in reference to the cross-collaborating that would ensue for a greater part of the conference. Before splitting up into assigned task groups though, participants participated in panel discussions and were attentive in listening to speakers. Quick to jump into question-and-answer mode, participants established an in-depth dialogue of topics early on. While considering serious issues in aging and healthspan with regard to scientific research and methods, there was a lot of laughter among participants, which created an environment of ease while promoting a place of intuitive, unrestricted thinking.

The conference keynote address was given by Michael Merzenich, professor of otolaryngology at the University of California, San Francisco. He broke down the topic of healthful longevity, an idea involving a longer life in a body that still works. Breaking down the cycle of life into epochs, Merzenich outlined our brain function from child to old age.

Intentionally diverse task groups met for the first time after lunch on day one of the conference. Meeting for roughly nine hours in total over two and a half days, each group tackled the challenges and questions presented to them and developed scientific plans. Groups explored topics such as

- the relationship between demographic mortality rates, aging, and functional human healthspan;
- the effects of exercise on human healthspan;
- the cellular and molecular mechanisms of biological aging;
- inflammation's effects on aging; and
- changes in social context to enhance functional status of the elderly.

Other groups were tasked with designing new research paradigms to assess healthspan, enhancement and prolongation in experimental research models, and developing technological interventions to overcome barriers to independence and community participation.

Thinking of where to even begin work on those topics was a challenge, so participants in each group decided they needed to be on the same playing field by defining key terms. Going into midconference group progress reports, participants came to the consensus that healthspan should be defined as the length of time an individual is able to maintain good health, but would not be equated with lifespan. Health was defined as the ability for a system to maintain or return to homeostasis in response to challenges. Because health is not quantifiable, many agreed that it was necessary to come up with a series of tests to somehow calculate it.

Some task groups initially had trouble getting off to a running start, but others eased into project design and discussion. Some members butted heads on ideas, while others smoothly worked off of one another to make giant leaps of progress. The nature of the task groups was greatly varied, but all had their sights set out for them as they paved the way toward solutions for their challenges.

Group members not only discussed solutions for the near future but also allowed themselves to imagine possibilities far into the future. Some floated around science-fictionlike ideas of new technologies for changing human behavior, limb and organ regeneration, and artificial intelligence.

All agreed that in the midst of all the development—from new technologies to assist aging or encouraging certain social behaviors in the elderly—the most important aspect was to ensure access to those services by establishing relationships with health care providers, as well as insurance providers.

Even at the end of each day, as task group members grew weary of the long hours spent thinking of ways to address challenges presented to them, they were still eager to network and relax with their newfound colleagues.

GETTING IT ACROSS

Communicating the depths and complexities of science topics takes great skill and practice—an investment that science and health care journalists make to get the facts straight. Thirteen journalism students from graduate science writing programs around the country were invited to cover this year's conference, a coveted opportunity. Each writer participated in a task group, some not just as reporters but also as active contributors to their group's discussions, and their summaries have been compiled to provide an overall picture of the conference.

The initiative stresses the importance of communication in its mission, which is why they recognize talent and excellent work each year. A selection committee, after an almost yearlong deliberation over hundreds of books, articles, and broadcast media, decided on the winners for this year's Communication Awards, which were presented at dinner on November 14th.

Dr. Eric Kandel, professor at Columbia University, won in the book category for his memoir, *In Search of Memory: The Emergence of a New Science of Mind*, written from a scientist's perspective about the human mind. Freelance writer Carl Zimmer won in the newspaper, magazine, and Internet category for his diverse coverage of evolution and biology. Among his works were "Devious Butterflies, Full-Throated Frogs, and Other Liars," published in the *New York Times*, and "A Fin Is a Limb Is a Wing," published in *National Geographic*. Zimmer also has a science blog hosted by *Seed* magazine called "The Loom." For the television and radio category, host and producer Jad Abumrad, cohost Robert Krulwich, and senior producer Ellen Horne were presented with an award for their work on Radio Lab's "Where Am I?" and "Musical Language," exemplary examples of making science accessible to general audiences.

AT THE END OF IT ALL

The time spent together was short, but participants came out with a newfound excitement in their research, probably the most important aspect of the conference. The task group challenges and questions acted as a ground for establishing those working relationships, in hopes of fostering future collaborations between members.

The initiative offers grants to all participants of the conference to encourage innovative thinking and new projects among those who choose to collaborate with one another. Applicants for these grants can get up to $100,000 for their suggested projects, and NAKFI plans to award up to $1 million for research related to human healthspan.

While the conference itself has passed, many researchers are well on their way to developing new ways to address aging. You never know, you might soon see robots as caretakers for humans, or more elderly citizens participating in more outdoor community activities.

Enhancing the Functional Status of the Future Elderly

TASK GROUP DESCRIPTION

Background

Disability rates in older persons for both activities of daily living (ADL) and instrumental activities of daily living (IADL) have been falling steadily for more than two decades. Whether these declines will continue or reverse is a major concern of considerable debate as experts argue how the current epidemic of obesity and diabetes among baby boomers and younger cohorts will affect the functional abilities of these generations. Coming decades will likely bring significant technological advances, both medical and nonmedical, that will importantly impact the capacity for the elderly to function. With respect to medical technologies, numerous technologies—from smart prosthetics to distance monitoring devices—will enhance function and independence. However, these technologies are likely to be expensive, and national policies regarding the amount of cost to be absorbed by patients may place many of them out of reach for those in the lowest socioeconomic group, further widening the functional gap between the haves and the have-nots in our society.

Nonmedical technologies, such as striking advances in computerization in the workplace and beyond, may either enhance the capacity of older persons to function or outstrip their capacity to adapt—thus aggravating the digital divide between generations. These nonmedical technologies also can enhance general functional status and independence, as can be seen in the increasing use of the Internet by older persons to pay bills, shop, and communicate with friends and family.

Initial Challenges to Consider

1. Develop projections for future age-specific disability rates by gender, race, and socioeconomic status. These projections should take into account:

- Various definitions of disability (e.g., ADL, IADL, cognitive function, other definitions and subtypes of disability);
- Recent trends in disability rates for various age groups;
- Expected future changes in lifestyle (e.g., diet, exercise, smoking), health care (new approaches to diagnosis and treatment), and health status (e.g., obesity, diabetes) that may influence disability;
- Changes in medical and related technologies that may compensate for functional impairments; and
- Advances in nonmedical technologies, in the workplace and beyond, that may enhance or limit function in older persons.

2. Develop new concepts for interventions (e.g., social, medical, technological) that mitigate these trends.

Initial References

Ahacic, K., L. Kåreholt, M. Thorslund, and M. G. Parker. 2007. Relationships between symptoms, physical capacity and activity limitations in 1992 and 2002. Aging Clinical and Experimental Research 19(3):187-193.

Bean, J. F., A. Bailey, D. K. Kiely, and S. G. Leveille. 2007. Do attitudes toward exercise vary with differences in mobility and disability status? A study among low-income seniors. Disability and Rehabilitation 29(15):1215-1220.

Cooper, R. A., M. L. Boninger, D. M. Spaeth, D. Ding, S. F. Suo, A. M. Koontz, S. G. Fitzgerald, R. Cooper, A. Kelleher, and D. M. Collins. 2006. Engineering better wheelchairs to enhance community participation. IEEE Transactions on Neural Systems and Rehabilitation Engineering 14(4):438-455.

Cooper, R. A., H. Ohnabe, and D. Hobso (eds.). 2006. *An Introduction to Rehabilitation Engineering,* 1st ed. London: Taylor & Francis.

Cutler, S. J. 2005. Technological change and aging. In *Handbook of Aging and the Social Sciences,* 6th ed., eds. R. H. Binstock and L. K. George, pp. 257-276. San Diego, Calif.: Academic Press.

Freedman, V. A., N. Hodgson, J. Lynn, B. C. Spillman, T. Waidmann, A. M. Wilkinson, and D. A. Wolf. 2006. Promoting declines in the prevalence of late-life disability: Comparisons of three potentially high-impact interventions. The Milbank Quarterly 84(3):493-520.

Motta, M., L. Ferlito, S. U. Magnolfi, E. Petruzzi, P. Pinzani, F. Malentacchi, I. Petruzzi, E. Bennati, and M. Malaguarnera. 2008. Cognitive and functional status in the extreme longevity. Archives of Gerontology and Geriatrics 46(2):245-252.

Task Group Members

- Liming Cai, National Center for Health Statistics/Centers for Disease Control and Prevention
- Vicki Freedman, University of Medicine and Dentistry, New Jersey
- Jong-in Hahm, Pennsylvania State University
- Brian Hofland, The Atlantic Philanthropies
- Jeffrey Kaye, Oregon Health and Science University
- Erin Lavik, Yale University
- Howard Leventhal, Rutgers University
- Graham J. McDougall, University of Texas at Austin
- Michael Merzenich, University of San Francisco
- Laura Mosqueda, University of California, Irvine
- Elaine Oran, U.S. Naval Research Laboratory
- Donald Royall, University of Texas at San Antonio
- Michael Tannebaum, University of Georgia

TASK GROUP SUMMARY

By Michael Tannebaum, Graduate Writing Student, University of Georgia

Defining the Problem

Disability rates in older persons for both activities of daily living (ADL) (e.g., bathing, dressing, and eating) and instrumental activities of daily living (IADL) (e.g., financial management, taking medications, and grocery shopping) have gradually declined over the last two decades. However, with both the baby boomer generation and younger cohorts currently experiencing increases in the prevalence of diabetes and obesity, as well as potentially increasing numbers of survivors living longer with chronic illness, declining disability rates are in jeopardy of coming to a halt and reversing.

Studies to date provide an evidence base for suggesting methods that may sustain function or prevent excess disability (e.g., Freedman et al., 2006). Nevertheless, at the individual level it is recognized that there are many challenges to implementing and maintaining interventions to promote independent living. Thus, although people may be aware of the behaviors that potentially contribute to a long, healthy life (e.g., physical activity and good dietary habits), both in the short and long term, many people nevertheless neglect the long-term consequences of their current lifestyle choices for a number of reasons. These reasons range from the pull of more immediate gains (e.g., plastic surgery) to lack of resources or practical, easy-to-follow knowledge of how to implement and continue programs of change in their lives. To enhance the functional status of the future elderly and ensure that disability rates continue to decrease, appropriate interventions must be developed and people need to be willing to adopt, embrace, and sustain the necessary lifestyle modifications.

The potential for applying existing knowledge with regard to maintaining function in later life is enhanced by the realization that in the coming decades, significant medical and nonmedical technological advances will importantly impact the capacity of the elderly to function independently. With respect to medical technologies, numerous technologies—from smart prosthetics to distance monitoring devices—will enhance function and independence. Existing nonmedical technologies are already increasing in use to enhance function by older persons. This can be seen in the growing use of the Internet by older persons to pay bills, shop, and communicate with friends and family. However, these technologies face several challenges. They need to be proven effective and integrated into the mainstream of functional intervention methods that are already known to be effective. Without this evidence base, their initial costs are unlikely to be covered by traditional medical models of reimbursement. However, to the degree that this new health-sustaining technology is piggy-backed as part of devices in use in everyday life (e.g., phones, televisions, common appliances) and their value translated clearly to the average consumer, the cost issues may not fall into conventional health care models. This is emphasized by the fact that nontraditional health-oriented companies and industries (e.g., Microsoft, Google, Intel, Wal-Mart) are rapidly moving into the health care arena, suggesting that access to these technologies across a wider segment of the socioeconomic spectrum is more likely to become realized. Finally, a unique aspect of employing new technology-enhanced approaches to interventions sustaining function is the potential for the technologies themselves to provide better feedback to the health care system with regard to how people are functioning. Thus, for example, if daily activity is captured unobtrusively and electronically in the home as part of a functional maintenance program, a more real-time record (as opposed to recall or diary methods) of how people are doing day-to-day may be achieved.

Charge to Task Group

This multidisciplinary group comprised 13 individuals trained in fields such as neuroscience, demography, cognitive science, epidemiology, geriatrics, and bioengineering. The group had two predominant challenges. The first was to develop projections for future age-specific disability rates by gender, socioeconomic status, and race, which also required the group to ascertain the data and measures necessary to generate these projections. The second was to identify novel concepts for social, medical, and technological interventions that would be capable of preventing further declines in disability. Consensus was reached that adapting the methods and outcome measures from the intervention studies should be employed to drive the data collection used to project future age-specific disability rates as discussed in the first charge.

The group hoped to formulate interventions that would successfully transform an individual's thinking from the "living for today" mindset to one that places emphasis on life's later years as well. To accomplish this the group set out to identify potentially efficacious incentives, which would encourage people to adopt and sustain desirable behaviors with respect to both the short and long term. Consensus among group members was that populations targeted for intervention need incentives in order to take up and continue the desired behavior or lifestyle modification. The group agreed that incentives need to be catered toward the individual, taking into account cultural and economic factors of the targeted population. In addition, the group believed that people would be more likely to begin and maintain a desirable behavior if potential barriers were minimized, which can in part be accomplished by ensuring interventions are cost-effective and attractive to the target population.

Strategy

Prior to discussing specific interventions, the group set out to identify key outcomes to change through the interventions. It was concluded that extensive variability exists in aging and that the ideal outcome for one person may be different from that for another. Therefore, the group determined that the desired outcome should capture this variability. Overall, maximizing functional status across multiple domains (e.g., physical, cognitive, and mental) and well-being as a function of life expectancy was deemed to be the key outcome or goal.

The group next framed important considerations for conducting interventions, such as which populations to target, the role technology should play, and the optimal time for intervention (i.e., intervening upstream vs. downstream). Intervening upstream means intervening at a relatively young age with hopes of changing a widely distributed risk factor by a small amount, accruing benefit across the lifespan but with a large impact at later ages. Because intervening upstream typically targets a wider population (unless specific risk factors are known at young ages), resources and time may be allocated where intervention is unnecessary or ineffective. Intervening downstream, on the other hand, means intervening later in life. The intervention may be more focused and precise; however, it may be too late and too expensive for the intervention to be effective. Since both upstream and downstream interventions have advantages and disadvantages, the group concluded that the point of intervention ought to be case specific and depend on factors such as the type of intervention employed, the desired outcome, and the target population.

Additional considerations discussed were the feasibility of intervening at different stages within the upstream or downstream construct (e.g., fetal development, preclinical, or clinical), unit or levels of analysis (e.g., cellular biomarker, individual, family, community, societal), and which predictors best explain the variance of outcomes. In addition, emphasis was placed on the need to develop interventions that allow members of the target population to maintain a degree of autonomy, control, and choice. Finally, the group explored the idea of using a combination of interventions and the potential to get multiple desired effects from one intervention. The latter approach has not been typically explored in current functional intervention research.

Since no standard blueprint exists detailing how to effectively combine intervention strategies, identify the exact populations to target, and determine the perfect point in the life course that an intervention should begin, the group decided that testing several integrated, multifactor, multilevel approaches would yield the best results. In addition, smaller-scale interventions that have proved successful in the past should be implemented in more fully realized versions to test their scalability.

Interventions

The group was aware of a recent workshop sponsored by the National Institute on Aging, the National Academies, and the National Research Council (Workshop on Identifying New Interventions to Extend Disability Decline in Elderly Populations, September 14-15, 2006) that explored interventions that may be applied to the aging population. Thus, they were interested in considering not only these methods, which were well known to the task force, but also other opportunities as well. The group came up with more than a dozen potential intervention ideas or concepts, which ranged from social or community-based interventions (e.g., social engagement initiatives and nutritional happy hours) to technology-driven ideas. Moreover, the possibility of establishing a reward system that encourages people to take care of themselves (e.g., a health advantage credit card) was explored.

Among the more biological interventions included a proposal to establish stem cell banks in which an individual's tissues are stored in case they are needed at a later point in that person's life. The more cells an individual has on file, the better chance he or she would have of regenerating tissues. A stem cell bank would facilitate new treatments such as cartilage injection instead of joint replacement. This technology and its potential are rapidly evolving; careful tracking of changes in this field relative to its application to prevent disability will be needed to apply over the long term.

Increased pervasiveness of technology in the household was a common theme among discussion of intervention ideas. The concept of harnessing the power of embedded or ubiquitous home computing was reviewed. In this paradigm a person might be monitored at home for subtle health and activity changes over time with, for example, motion-activity sensors assessing physical activity or sleep and when a high-risk trend-line was observed, an intervention might be applied or modified appropriately. One group member suggested a brain "gym" in which there is ongoing brain-based training, which would be largely self-administered and achievable through technological strategies. It was suggested that intensive computer-based training might eventually sustain the brain to mitigate the effects of pathologies of brain aging and neurodegenerative diseases.

Several group members pointed to the challenges of technology-enhanced interventions. For example, one group member argued that for some applications of technology one needed to take into account the high prevalence of executive control dysfunction in the elderly, which affects an individual's self-awareness and ability to use technology (although this fact may by itself suggest ways of detecting functional decline).

The group discussed many study designs that would incorporate the interventions reviewed. The concept of focusing on periods or groups at relatively high risk for decline was emphasized for immediate-term studies in order to increase the efficiency of some trials. For example, a study may choose to target intervention within an age group where the disability rate takes off (e.g., eighth decade) to more quickly observe whether the intervention effectively reduces disability or institutionalization. Another approach might be to pick people with one existing major ADL/functional dependency putting them at risk for further loss, and then conduct assessments (e.g., walking speed), to determine whether an intervention may reverse the decline. As noted above, the design of intervention studies must be tailored to the intervention and the outcome measures to be assessed.

A New Initiative

After considering many potential interventions designed to enhance the functional status of the future elderly, the group developed an action plan to move the field forward. Inherent in this plan was the notion that no single approach or method could be advocated at this time. However, there was ample evidence that building on existing interventions, using a holistic or multidimensional approach was most needed. In this context a national initiative to create and test additional interventions was proposed. The 10,000-person initiative, named the Studies of Optimal Functional Interventions in Aging (SOFIA), would call for a national initiative or competition to study in parallel within this population, using standard metrics, how several interventions might best sustain independent function among the elderly. The parallel studies would be complementary and likely be of various lengths, both short term (e.g., three months for testing novel methodologies) and long term covering different periods of risk. Within the studies it would be important that milestones are shown and reported early as well as trajectories of change over the course of the studies (not only simple monotonic outcomes). Technology that inherently measures trajectories of functional change would be embedded in all participants' homes. The research teams would receive a predetermined amount of money and would target a set or standard population with their intervention(s). This initiative would ideally be funded by a consortium of sources, including but not limited to the U.S. federal government, various industries (e.g., technology and care providers), private foundations, and prize money (possibly awarded by industry).

The Future

The group concluded that a meeting of leaders and stakeholders should be planned to establish ground rules for the aforementioned initiative. At this meeting the following topics (and many others) would likely be discussed:

- Identification of participants and their roles;
- Identification of plans for recruitment and retention;
- Determination of how immediate and long-term benefits and costs will be evaluated;
- Data collection, analysis, sharing, and feedback;
- Optimal timing for intervention;
- Study period and funding cycle;
- Scalability (make sure studies don't just work for 100 people, but for 1,000); and
- Dissemination and sharing of results.

It is the hope of this group that the results from this initiative will also function as a national assessment of disability change over time, thus helping to develop projections of future age-specific disability rates according to various demographic, economic, and behavioral factors.

Although the current obesity and diabetes epidemics among baby boomers and younger cohorts threaten to increase disability rates in the future elderly, effective interventions applying established and new methodologies and technology have the potential to not only forestall an increase in disability rates but also to improve these trajectories of change for future generations.

Design New Research Paradigms to Assess Healthspan, Its Enhancement, and Prolongation in Experimental Research Animals

TASK GROUP DESCRIPTION

Background

Enhancing and prolonging human health is a worthy social goal as specifically expressed in the mission statement of the U.S. National Institutes of Health. Indeed, a remarkable decline in disability among people 65 and older in the United States has been thoroughly documented since 1982 (Manton et al., 2006). By contrast, increasing lifespan without prolonging healthspan could easily be viewed as a societal catastrophe, necessitating an ever-growing fraction of national resources to be devoted to the treatment and care of the disabled elderly.

Basic biological researchers, fascinated by the prospect of altering the rate of aging, have been remarkably successful over the past two decades in discovering genetic, pharmaceutical, and environmental treatments that increase the lifespan of laboratory research animals. Specifically, laboratory longevity of the roundworm, *Caenorhabditis elegans*, has been experimentally increased more than sixfold; the fruit fly, *Drosophila melanogaster*, more than twofold; and the house mouse, *Mus musculus*, by as much as 75 percent (Bartke et al., 2001; Partridge and Gems, 2002). However, aside from a handful of studies of fruit flies (Burger and Promislow, 2006; Van Voorhies et al., 2006), comparatively little effort has gone toward determining whether these longevity-extending treatments also enhance and prolong animal health and functional capacities (Bartke, 2005). Some evidence suggests that experimentally enhanced longevity may have deleterious early-life consequences (Jenkins et al., 2004) and even the most extensively documented life- and health-extending treatment (calorie restriction) may increase susceptibility to some common infectious diseases (Gardner, 2005). Furthermore, treatments such as chronic exercise that appear to enhance health but have negligible impact on lifespan have received little interest or attention from biogerontologists (Holloszy, 1997).

The aim of this task group is to provide strategies for assessing the lifelong health and functional capacities of animals traditionally used in aging research and thus to aid in the discovery of new medical treatments that improve health and healthspan irrespective of their effects on lifespan.

Initial Issues to Consider

- How should we define health and healthspan? Are they simply the lack of specific disabilities or should they encompass positive measures of functionality?
- To what extent can the typical laboratory environment with its superabundant food; its constant, benign, pathogen-defined environment; and its limitations on physical activity allow the assessment of animal health and well-being?
- How might the living environment of the laboratory be altered to reveal more about animal health?
- In addition to designing a living environment that is more revealing about animal health, would periodic challenges be useful to assess cognitive, sensory, and physical capacities as well?
- How much can we infer about animal health from demographic information alone?
- Would alternative animal models allow better assessment of health measures relevant for humans?

Initial References

Bartke, A. 2005. Minireview: Role of the growth hormone/insulin-like growth factor system in mammalian aging. Endocrinology 146(9):3718-3723.

Bartke, A., J. C. Wright, J. A. Mattison, D. K. Ingram, R. A. Miller, and G. S. Roth. 2001. Extending the lifespan of long-lived mice. Nature 414(6862):412.

Burger, J. M., and D. E. Promislow. 2006. Are functional and demographic senescence genetically independent? Experimental Gerontology 41(11):1108-1116.

Gardner, E. M. 2005. Caloric restriction decreases survival of aged mice in response to primary influenza infection. Journals of Gerontology A—Biological and Medical Sciences 60(6):688-694.

Holloszy, J. O. 1997. Mortality rate and longevity of food-restricted exercising male rats: A reevaluation. Journal of Applied Physiology 82(2):399-403.

Jenkins, N. L., G. McColl, and G. J. Lithgow. 2004. Fitness cost of extended lifespan in *Caenorhabditis elegans*. Proceedings: Biological Sciences 271(1556):2523-2526.

Manton, K. G., X. Gu, and V. L. Lamb. 2006. Change in chronic disability from 1982 to 2004/2005 as measured by long-term changes in function and health in the U.S. elderly population. Proceedings of the National Academy of Sciences U.S.A. 103(48):18374-18379.

Partridge, L., and D. Gems. 2002. Mechanisms of ageing: Public or private? Nature Reviews Genetics 3(3):165-175.

Van Voorhies, W. A., J. W. Curtsinger, and M. R. Rose. 2006. Do longevity mutants always show trade-offs? Experimental Gerontology 41(10):1055-1058.

Due to the popularity of this topic, two groups explored this subject. Please be sure to review the second write-up, which immediately follows this one.

TASK GROUP MEMBERS—GROUP A

- Kath Bogie, Case Western Reserve University
- Daofen Chen, National Institutes of Health, National Institute of Neurological Disorders and Stroke Neuroscience Center
- Matt Kaeberlein, University of Washington
- Karim Nader, McGill University
- Corinna Ross, University of Texas Health Science Center
- Richard Sprott, Ellison Medical Foundation
- Heidi A. Tissenbaum, University of Massachusetts Medical School
- John Cannon, University of California, Santa Cruz

TASK GROUP SUMMARY—GROUP A

*By John Cannon, Graduate Science Writing Student,
University of California, Santa Cruz*

Richard Foster, board member of the W. M. Keck Foundation and managing partner of the Millbrook Management Group LLC, drew inspiration from the life of William Keck during his opening address to the 2007 National Academies Keck *Futures Initiative* conference. As a wildcatter working on the oil rigs in early 20th-century California, Keck eventually started the Superior Oil Company, and was said to have drilled 23 dry wells before striking "black gold" on his 24th try.

The idea was not to drill dry wells, Foster said, but calculated risk in the spirit of Keck's innovative thinking can have great potential.

"Think big, risky projects," Foster said.

And that's exactly what this task group did when charged with addressing healthspan research in animal models. Instead of examining just a few specific models, then delineating their respective pros and cons and proposing a set of recommendations, the group decided it would be more beneficial in the long run to invest in a physical and virtual center—a repository of available models where scientists could go and learn the best strategies for answering their questions of interest.

Establishing the Problem

Representing engineering, biology, and neuroscience, the group of seven researchers each brought a different set of experiences in the realm of animal models. Collectively they agreed that huge gains had been made in extending the lifespan of animals. More difficult was deciding whether those gains included correlated advances in healthspan as well.

"Is lifespan not a sufficient proxy [for healthspan]?" said Matt Kaeberlein, assistant professor of pathology at the University of Washington. To answer this question the group needed to hammer out ways to dissociate longevity from the process of healthy aging.

An early suggestion was to search for ways to increase the healthspan of an animal without increasing its lifespan. Only then would the investigator understand the effect of whatever she had chosen to manipulate, according to that line of thinking.

Michael Rose, professor of ecology and evolutionary biology at the University of California, Irvine, and a floater during the conference, thought it might be more productive to steer the group in a slightly different direction.

"That's a really cool idea, but as someone who does this for a living, it would be really hard," Rose said. Instead, he proposed it would be more straightforward to find attributes that increase both lifespan and healthspan. "We have selected on characters that are initially correlated with lifespan, like acute stress resistance, for example. We have generally found that when you select for increased late-life function, then you find almost all your measures of what you consider healthy for the organism increase."

At issue here was a definition of health, something much more difficult to measure than the days, months, or years that can be added to life.

And so a working definition of health began to materialize: the ability of a system to maintain or return to homeostasis in response to challenges. It wasn't perfect and didn't encompass every aspect.

But the definition provided the group with a starting point to answer the question of how to quantify health. Then, by extension, healthspan would be how long an individual could maintain good health.

"If that's our definition of healthspan, how are we actually going to test it?" asked Heidi Tissenbaum, associate professor at the University of Massachusetts Medical School.

In keeping with the intended focus of the task group, the discussion turned to differentiating what makes certain animal models good candidates for measuring healthspan.

There has been a tendency to stick to a few well-known animal models when it comes to aging research, said Corinna Ross, a primate behaviorist and postdoctoral fellow at the University of Texas Health Science Center at San Antonio. "We already know all kinds of things about mice, but I don't think there is enough thinking 'outside the box.' Even bringing wild mice into the lab was a completely unusual thing."

"Maybe one of the suggestions a group like this could provide is the development of a center for alternative models," said Richard Sprott, a behavior geneticist and executive director of the Ellison Medical Foundation.

Development of the Idea

Choosing one specific animal to research can be a daunting task. Each model has its advantages and drawbacks. For example, the worm *C. elegans* is a common model in aging and other types of research. Scientists have been able to increase its longevity at least sixfold, but it also can enter a dauer stage in which the organism doesn't age—something that complicates extrapolating findings to humans. So, although there's a lot we can learn from this quirky nematode, it isn't a perfect model, just as no model is perfect.

So too, members of the group expressed concern that researchers might be limited by their own institutions. Say, for instance, that a researcher is interested in a set of questions dealing with aging and does find an "ideal" animal model to begin researching. That's just the first step. Perhaps the researcher's institution doesn't have the facilities to accommodate that particular species. Or maybe the institution's leadership has a decided bias toward an entirely different model, one that doesn't have the potential of the ideal organism. In both cases the only solution would be to leave that stone unturned and move on to another question, perhaps losing the opportunity to contribute significantly to the field's base of knowledge.

Kath Bogie, senior research associate in orthopedics at Case Western Reserve University, was struggling with just those sorts of troubling issues. Her investigations into ischemic wound healing in rabbits hadn't quite produced the results she had hoped for, and she asked for advice about other available animal models.

As the discussion ensued, the proposed center seemed as though it would be an excellent resource to come to the aid of researchers with questions similar to Bogie's.

The particulars of any idea with such a large scale are always tricky, but there was no devil in these details. The collective experience of the group came together, jettisoning this big idea into a plausible innovation. Central to the mission of the institution should be to shed light on the models that have the most potential to answer a particular question about aging. Further, the center shouldn't address only which animal model, the group concluded, but also which challenges might yield the most telling results, given the specific question.

Not long into the discussion did the issue of money arise.

"How much would a center like this cost?" asked Karim Nader, associate professor of psychology at McGill University.

Richard Sprott, citing a past venture with a similar mission, speculated that with a startup cost of $10 million, the center could get off the ground with an annual budget of between $5 million and $7 million. But steps should be taken so that it would be able to weather the storms of fickle funding sources.

"As a practical matter it would require a long-term business plan to describe how it gets to self-sufficiency in something like a 10-year span," Sprott said.

Proposed Solution

As the second day of discussions drew to a close, the group assembled a presentation justifying the need for such a center and detailing what they saw as the next step in the process of its inception.

The strengths and limitations of the few most popular animal models are well known, but the center would allow for the exploration of alternative models—ones that could help address specific problems more precisely—with the intent to make them available to researchers who are involved with the center. In the spirit of the NAKFI conference and the free exchange of ideas, the center would have an open-access capacity so that researchers would have the best access possible to available models. In doing so, this would also create interdisciplinary cooperation. Perhaps most importantly the center would be established to advise researchers on the most appropriate models for their area of interest in healthspan research.

The Next Step

To further this idea the group decided a future meeting should be convened. Invitees should include prominent figures in the aging research community, as well as representatives of potential funding sources and from the National Academies to discuss more specifically what models and assays would be included in the initial development of the center. They would also be charged with drumming up support from the broader aging research community and hammering out the initial details of what the actual center would entail. Among the questions they would address are: What will be its physical presence? How will virtual models be incorporated into the center's design? What are the best ways to encourage the participation of investigators?

"In talking to a few people about what our group had developed, I got two responses. One was skepticism," Kaeberlein said at the close of the conference. "There's reason to be skeptical. But the other response, which even the skeptics had, was unanimous support for the idea that aging research is ready for this idea for this type of center. Maybe this isn't the perfect model, but I think it's a great place to start."

TASK GROUP MEMBERS—GROUP B

- Allyson Bennett, Wake Forest University
- James Carey, University of California, Davis
- James Herndon, Emory University
- Lauren Gerard Koch, University of Michigan
- Sean Leng, Johns Hopkins University
- Daniel Perry, Alliance for Aging Research
- Shane Rea, University of Colorado
- Felipe Sierra, National Institute on Aging

- David Waters, Purdue University
- Molly Webster, New York University

TASK GROUP SUMMARY—GROUP B

By Molly Webster, Science Writing Graduate Student, Science, Health, and Environmental Reporting Program, New York University

Traditionally, aging has been studied as a finite period in an organism's life history. Our task group, however, quickly concluded that this is the wrong way to look at the event. An organism does not go from healthy to dead: It moves through a process that shuttles it from the first point to the last. Therefore, aging should be studied throughout a lifespan (defined as birth until death). Additionally, aging is not only a life course event, it has also been shown to be multigenerational. An ancestral generation's genetic makeup or its socioeconomic parameters will affect how a subsequent generation ages. To account for these new ideas the task group decided to focus on studying longevity in animal models, using a continuous, transgenerational approach.

Declining health is a natural part of aging, but it became apparent that though health is something we can abstractly understand, varying perceptions make concretely defining it tricky. Few 14-year-olds will think a 60-year-old healthy, but the elder feels he is fitter than a 52-year-old heart attack victim. Similarly, according to a task group member, many physically disabled patients consider themselves healthy, even though the nondisabled would not. The definition of health, and subsequently healthspan (the time of life before the frailty period), obviously varies per person and per demographic group. Therefore, we concluded that health cannot be defined. Rather, it needs to be established (1) per species, (2) by individual investigators, and (3) per particular environment. Having established these more general guidelines, the task group moved on to consider the reality of an ideal animal model. Answering this question helped us then talk about new research approaches for studying aging as we defined it. Lastly, we discussed the merits of benign lab environments.

The Arc of Aging

Plotting the aging life history of lab animals will be critical to understanding aging in every model species. This plan means following and documenting the aging arc of a lab model from birth until death. There are at least three ways to document this arc: (1) behaviorally, (2) physiologically, and (3) cognitively. Naturally, when plotting the changes an organism undergoes related to aging, longevity phases would emerge. Already it's assumed aging can at least be broken down into periods, such as youth, middle age, old age, and frailty, but these divisions are as of yet vague and uncharacterized, and perhaps more will emerge. Clearer definitions of these periods should allow researchers to locate quantitative points where an organism noticeably shifts from one phase to the next. We called these points thresholds, or endpoints, and we imagined them analogous to quantitative points physicians currently use to diagnose, such as blood pressure. Just as anything above 140:90 is considered high blood pressure, while anything below is healthy, once we've plotted the life history of a species, we will come to know that anything below x is expressive of one longevity phase, while above it is indicative of another.

The two age phases that the task group primarily focused on were death and frailty. What constitutes frailty will differ for every type of animal model, but based on currently available data in human studies, researchers can use five parameters to help judge a model's health:

- Exhaustion;
- Muscle strength (measured by grip strength);
- Weight loss;
- Walking speed; and
- Exercise or physical activity level.

For example, entomologists observe that fruit flies tend to flip over on their backs when they are not doing well or are frail. Researchers working with *C. elegans* also notice changes in physical appearance and movement trajectories of worms when they become frail. It is clear then that there are signs of frailty that are readily observable in animals and thus can be used to define frailty in a species. It was suggested that animal models with long frailty periods—and perhaps our lab rats or female organisms already express this—would be a good model choice, because this longer frailty period would allow for multiple different experiments. Death was tenuously proposed as a phase of aging at first; presently it is not coupled with aging. But seeing as death is a process (and it is a process) that normally follows frailty, we thought that it would be important to clearly understand, especially to determine the threshold between it and frailty. The group realized that studying death is an enterprising area of research that is currently not a viable research avenue. "We don't get to let our animals live until they die," said one group member. Instead, as most researchers are aware, institutional animal care and use committee (IACUC) regulations require euthanasia of lab animals before this life event. IACUC regulations must be changed, the group determined. The best way for this to happen is for the National Academy of Sciences to endorse studies of the dying process, and to propose regulations that would allow this to happen. The task group felt the weight of the Academy is the only way to change existing regulations.

Once we establish a baseline diagram of what aging looks like in a specific organism, we can then manipulate the organism and see how that affects longevity. Perhaps we can turn a gene off during one life stage, and watch to see how that changes the process of getting older. An important part of tracking aging in lab animals is remembering that aging takes place across an entire organism; we would want to see its progression at every level of function, particularly the tissue level. Perhaps the liver ages more slowly than the skin. If so, why? It would behoove researchers to know whether the liver possesses a molecular attribute that increases its healthspan and/or lifespan. If we look only at one organ, or one element of an organism, we will limit our knowledge and potentially miss a piece of the puzzle. Not only that, if we minutely characterize aging, perhaps scientists or medical experts will be able to treat it prophylactically. If we can understand how illnesses like diabetes and aging work together at a molecular level, for example, there's a chance we could predict the disease event before an organism starts to express it symptomatically.

"*C. Elegans* Need Not Apply"

When we talk about studying longevity in animal models, what is the ideal animal model for these experiments? Our group, whose members represented *C. elegans*, fruit flies, lab mice, rats, and dogs, as well as labs that use monkeys and humans, agreed that there is no ideal longevity animal model; every species offers something unique. For example, if we are trying to understand how socioeconomic factors affect aging, it was suggested that dogs would be the ideal animal model, for they live in the socioeconomic environment of their owner. If a researcher was interested in a transgenerational study, however, fruit flies would be preferred to dogs as they have higher fecundity rates. The research question should drive what model is used in the experiment.

DESIGN NEW RESEARCH PARADIGMS

Also, when determining the appropriate animal model, researchers should be encouraged to look beyond those ordinarily used, in an effort to gather as much information about aging as we can. Group members suggest studying wild-type species, such as porcupines, gerbils, and butterflies—all of which have shown extended longevity in the field. Someone also pointed out that if dogs were a resource, perhaps other pets, such as hamsters, could be as well. However, the complaint is that right now research funding is so tight, finding monetary support for any experiment is problematic, let alone one proposing we study aging in an unknown animal species. The group recommends that this needs to change, describing that there should be a sanctioned search for a new animal model: a call to experimentation that can be described as "*C. elegans* need not apply."

Once the appropriate animal model is chosen for an aging study, it can be used by researchers in two different ways: Either it will help establish the arc of aging in the ways previously described, or it will be employed to test hypotheses. The hypotheses models would undergo manipulations, ranging from environmental to genetic, and then researchers would see how these affect the established aging process. Interventions can also be done on the hypotheses model: We can use organisms to test reversing aspects of aging. If turning off a gene causes z, can we use a model to see whether something can be done to reverse z? Or if z happens naturally, a researcher could use animal models as a method for learning how to prevent the event.

Environmental Factors

One of the questions posed to this task group was whether or not the benign lab environment was a sufficient setting for aging research. It's obvious to any researchers that lab settings are not a great mimic of the natural world: The environment is sterile, food is abundant, and there are no challenges or stresses posed to the organism. While the group didn't think lab environments should change—they are necessary to cancel out experimental variables—there are ways we can improve the data obtained from them. For starters, the task group decided any promising longevity experimental results should be further tested in different, nonbenign environments, with the model undergoing different advantages and disadvantages. So, if a type of *Drosophila,* with a genetic manipulation, seems to be living longer than average, this fly should then be brought into a different lab setting to see how its lifespan plays out under different conditions. Does it still express a longer than average longevity if population density increases? Or if food becomes scarce? Along with this, we also concluded that challenges, or stresses, should become part of the lab environment. As one group member pointed out, when a doctor is trying to determine what members of a group of males have a bad heart, he doesn't take their resting EKG; he throws them onto a treadmill to see how their heart acts under stress, and how long it takes for it to return to homeostasis. A similar stress test should be done with lab animals; lab mice should see a cat every once in awhile, especially if they are showing promising results in aging studies. Just as stress is one of the best determining factors for heart conditions, stresses are critical for better, more thorough lab experiments.

Public Sentiment

The point was raised that informal comment sessions have shown the public isn't particularly keen to live longer. In our discussion we assumed that that's because they assume a longer life will also include a longer frailty period. But we believe that increased healthspan is part and parcel of aging studies; there is no point in having humans live longer if they are incapacitated. While scientists understand where the lay audience is coming from, they fear that their concerns will negatively affect aging research. There was also consensus that while scientists at some point need to make it clear—without overextrapolating—how their research in fruit flies relates to humans, it is not up to them to market their research to lay audiences.

Effects of Exercise on Human Healthspan

TASK GROUP DESCRIPTION

Background

It is well recognized that motion and exercise have substantial benefits in increasing human health, cognitive and physical function, quality of life, and longevity. However, the dose-response effects of exercise on improving certain systems and increasing the lifespan and healthspan have not been well elucidated. Given the aging populations, there is an increasing interest in lifestyle factors and interventions that will enhance the cognitive and physical vitality of older adults and reduce the risk for age-related neurological and functional disorders, such as Alzheimer's disease and hip fractures. But the knowledge concerning the applicability and the effect of moderate- to high-intensity exercise programs is very limited for older people with or without apparent cognitive and physical impairments.

Initial Challenges to Consider

- What are the social, psychological, cognitive, physical, and physiological benefits of exercise in increasing the human healthspan? It is important to establish an age-related dose response curve. What should be a desirable goal for a healthy 70-, 80-, or 90-year-old?
- What are the benefits of exercise in the prevention of injury resulting from lack of musculoskeletal coordination and cognitive deficits?

- How to develop sensitive outcome measures of exercise over inactivity at the molecular, cellular, tissue, and functional levels to show positive effects of exercise?
- How to develop experimental models for use by the human population?
- Given that physical activity is an inexpensive treatment that could have substantial preventive and restorative benefits for cognitive and brain function, how can such interventions be studied and how can they be implemented on a population-based scale?
- How will technology (e.g., novel equipment, improved environment) facilitate and motivate people to exercise on a regular basis? Specifics should include how technological innovation will make exercise more attractive and compelling for cardiovascular and musculoskeletal systems improvement (e.g., aerobic and resistance exercise).
- How can we change community design in rural, suburban, and urban settings to facilitate and encourage as well as promote exercise in younger and older populations? What are the recommended strategies for such enhancement?

Initial References

Bourdel-Marchasson, I., M. Biran, P. Dehail, T. Traissac, F. Muller, J. Jenn, G. Raffard, J. M. Franconi, and E. Thiaudiere. 2007. Muscle phosphocreatine post-exercise recovery rate is related to functional evaluation in hospitalized and community-living older people. Journal of Nutrition and Health and Aging 11(3):215-221.

Buckwalter, J. A., V. M. Goldberg, and S. L-Y Woo (eds.). 1993. Musculoskeletal tissue aging: Impact on mobility. Chicago, Ill: American Academy of Orthopaedic Surgeons.

Hardy, S. E., and T. M. Gill. 2005. Factors associated with recovery of independence among newly disabled older persons. Archives of Internal Medicine 165(1):106-112.

Judge, J. O., C. Lindsey, M. Underwood, and D. Winsemius. 1993. Balance improvements in older women: Effects of exercise training. Physical Therapy 73(4):254-262.

Kramer, A. F., S. J. Colcombe, E. McAuley, P. E. Scalf, and K. I. Erickson. 2005. Fitness, aging and neurocognitive function. Neurobiology of Aging 26(suppl. 1):124-127.

Littbrand, H., E. Rosendahl, N. Lindelöf, L. Lundin-Olsson, Y. Gustafson, and L. Nyberg. 2006. A high intensity functional weight-bearing exercise program for older people dependent in activities of daily living and living in residential care facilities: Evaluation of the applicability with focus on cognitive function. Physical Therapy 86(4):489-498.

Mangione, K. K., and K. M. Palombaro. 2005. Exercise prescription for a patient 3 months after hip fracture. Physical Therapy 85(7):676-687.

Marottoli, R. A., H. Allore, K. L. B. Araujo, P. H. Van Ness, L. P. Iannone, D. Acampora, P. Charpentier, and P. Peduzzi. 2007. A randomized trial of a physical conditioning program to enhance the driving performance of older persons. Journal of General Internal Medicine 22:590-597.

Melov, S., M. A. Tarnopolsky, K. Beckman, K. Felkey, and A. Hubbard. 2007. Resistance exercise reverses aging in human skeletal muscle. PLoS ONE 2(5):e465 (www.plosone.org).

Messier, S. P., R. F. Loeser, M. N. Mitchell, G. Valle, T. P. Morgan, W. J. Rejeski, and W. H. Etinger. 2000. Exercise and weight loss in obese older adults with knee osteoarthritis: A preliminary study. Journal of the American Geriatric Society 48(9):1062-1072.

Wu, S. C., S. Y. Leu, and C. Y. Li. 1999. Incidence of and predictors for chronic disability in activities of daily living among older people in Taiwan. Journal of the American Geriatric Society 47(9):1082-1086.

Task Group Members

- Albert Banes, North Carolina State University
- Bambi Brewer, University of Pittsburgh
- Nadeen Chahine, Lawrence Livermore National Laboratory
- Nandini Deshpande, University of Kansas Medical Center
- Stephen Intille, Massachusetts Institute of Technology
- Robert Jaeger, National Science Foundation
- Richard Macko, University of Maryland School of Medicine
- Charlotte A. Tate, University of Illinois at Chicago
- Geoffrey Graybeal, University of Georgia

TASK GROUP SUMMARY

By Geoffrey M. Graybeal, Graduate Writing Student, Grady College of Journalism and Mass Communication, University of Georgia

IRVINE, CALIF.—An interdisciplinary group of scientists and researchers has written a prescription to improve the human healthspan: Exercise.

"Exercise as medicine" was one of several key ideas discussed by this task group at the National Academies Keck *Futures Initiative's The Future of Human Healthspan* conference.

During November 14-16, 2007, at the Arnold and Mabel Beckman Center of the National Academies, this particular task group spent two days examining the topic of the effects of exercise on human healthspan.

The group concluded that exercise could follow the pharmaceutical model in which doctors prescribe doses, such as amount and frequencies, in order to promote better health and combat diseases.

"Exercise is a pill. There's dose intensity to it," said Dr. Richard Macko, an associate researcher for the Baltimore Geriatric Research, Education, and Clinical Center and associate professor in the departments of neurology and gerontology at the University of Maryland and the Baltimore Veterans Affairs Medical Center. Macko served as chairman of the task group.

The practice gap between the medical community and the health community is huge, Macko told group members, who wondered whether a physician-recommended prescription of exercise would affect exercise results. For stroke victims there is evidence that a physician's recommendation affects patient efficacy, Macko told his colleagues in the group.

The exercise prescription could positively alter the health care of patients with chronic diseases, such as diabetes and hypertension. Group members discussed the importance of linking the medical requirement of exercise as a form of medical care, and also to provide continuity between the medical community, private sector, and government agencies and community advocacy organizations.

Dr. Charlotte ("Toby") Tate, dean of the College of Applied Health Science at the University of Illinois at Chicago, noted that the American College of Sports Medicine and the American Medical Association recently started an Exercise is Medicine™ initiative. One problem with that new program has been that primary care doctors typically haven't had exercise or nutrition training in medical school, Tate said. This raises the question of how to provide the desired continuity between the health and medical sectors, according to Tate.

Group members wanted to explore the possibility of expanding electronic medical records so that they could be accessed from any location, and also having exercise efforts become a performance measure that could be tracked. "If these recommendations don't have teeth they're going to be wallflowers," remarked one member.

The medical community should make exercise as medicine more pervasive than giving pills, said John Doyle, of the California Institute of Technology, a floater, who spent a great deal of time with this task group. The biggest problem, or inadequacy, however, he noted is what engineers call system integration. The drug industry spends billions of dollars on advertising to get the drugs into people's hands. A similar, across-the-board "full-court press" is needed for prescribing exercise, Doyle said.

The group generally agreed that exercise should be treated as a medi-

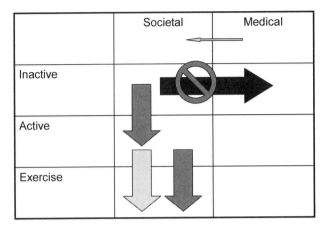

FIGURE 1 The grand integration challenge.

cine, prescribed and tracked and efforts to promote exercise moved into the community and corporate sector. The group largely agreed that technology use could encourage exercise activities and that business could play a role in supporting research initiatives.

Group members would like for the societal model to be linked with the medical model and a socioeconomic model also explored (see Figure 1 for an illustration of this grand integration challenge). "Is exercise important regardless of any of the other issues?" was a question that was posed. The evidence is yes, the group said.

In addition to prescribing the exercise-as-medicine approach, members of the task group would also like to see more people join the dance (dance) revolution, work toward a "SustainableYou," get a "SecondLife with exercise in it," go green to reduce their carbon footprint, and see more displays of "PDA." These were among the more creative solutions bandied about by members of this task group.

No, the researchers don't want to see public displays of affection. Nor do they believe personal digital assistants like Blackberrys or Treos are the answer either. The PDA the group envisions are "Purpose Driven Activities," a term suggested by Dr. Al Banes, a biomedical engineering researcher with a joint appointment to the University of North Carolina and North Carolina State University. The activities could include encouraging people to walk to school, work, and house of worship and organizations could partner with existing community efforts, such as the Rails-to-Trails organization that converts abandoned railroad trestles into walking paths.

One idea tossed around was a community program that would pair senior citizens with young children, where both would walk to school. One member noted that walking to school, rather than busing would be a great idea and that with gasoline prices continually escalating, this could be a forced reality.

The "SustainableYou" idea was proposed by Doyle as a campaign to connect physical activity to the greater issue of sustainability and global warming as a means of reducing an individual's carbon footprint.

The group discussed ways to encourage movement among kids and adults, ways to integrate movement into their everyday lives. The Nintendo Wii home video gaming system, in which users physically swing the controller, and Dance Dance Revolution arcade game, in which users hop from square to square, were cited as appealing means of technology that encourage physical activity but at the same time are popular and widely used. The group considered what other sort of technologies could be used in a similar fashion to tap into the public's competitive nature and love of games. Virtual reality programs that would allow a variety of physical challenges in a competitive environment were considered.

"We need a Star Trek holodeck," Banes said, referencing the simulator on the futuristic spaceship in the popular science fiction television series.

"We need a SecondLife with exercise in it," Doyle later added, referring to a popular online virtual community.

Business can be a powerful influencer of change, it was also noted. Some insurance companies use pedometers to track movement and if customers are participating they receive lower rates. The question of whether insurance companies and business provide enough incentives to get people to comply with good health was raised.

Group members also would like to call for the creation of a new exercise-monitoring device that will give positive feedback to encourage people to continue exercising. Such a device could include features such as a cell phone/wrist watch energy expenditure monitor, a SustainableYou carbon count monitor, and a dietary adviser that could help with meal suggestions, estimation of caloric intake, etc.

Problems

Before coming up with these possible solutions, the group grappled with meta-issues related to its task. The earlier discussion focused on the distinction between exercise and physical activity, the issue of whether the group should focus its efforts on exercise activities for disabled populations, the general populace, or both; how to treat technology initiatives; and an operational definition of what exactly "compliance" means.

Throughout the discussion the group realized that much of what is needed is outside of the realm of the hard sciences. A large part of the discussion centered on social science issues, such as how to motivate people to act, how to disseminate information and campaigns to the public, and the importance of moving efforts out of the lab and into the community. Group consensus was that moving beyond RCTs (randomized controlled trials) requires partnership and buy-in with community groups, businesses, and perhaps in some instances, with insurance providers. Trials for every form of exercise are not practical, a group member noted. Finding ways to motivate the public to exercise was a continual focus of the conversations.

Stephen Intille, technology director of the House_n Consortium in the MIT Department of Architecture suggested that technology might be used in innovative ways to ease people into exercise, diminishing the mental and physiological barriers to getting started. Many sedentary activities are instantly fun and engaging, such as watching a television show, snacking, or reading a book, while the payoff from exercise is typically far more delayed and subtle (as illustrated in Figures 2-6).

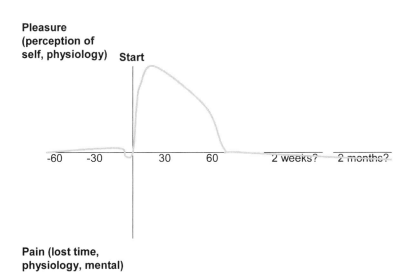

FIGURE 2 Ease of access to food.

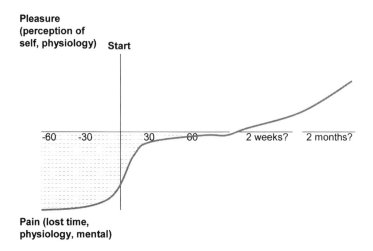

FIGURE 3 Exercise.

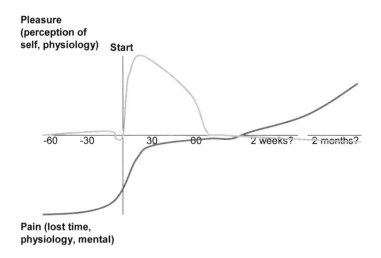

FIGURE 4 Perfect story (+ sedentarism).

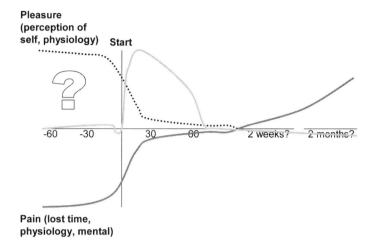

FIGURE 5 Research challenge: Fill compliance void.

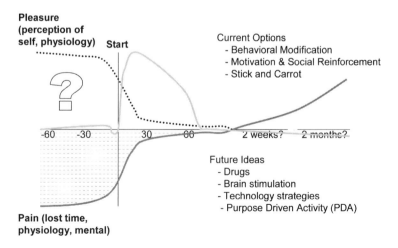

FIGURE 6 Research challenge: Get compliance/offset pain.

The author (Graybeal) noted that health communication is a bourgeoning field that receives a fair amount of funding from government organizations like the Centers for Disease Control and Prevention. Health communication professionals design campaigns aimed at communicating health initiatives and messages to the public. Graybeal suggested that researchers could partner with health communicators to disseminate their research by developing programs and messages to the public at large and specific community groups in particular.

Some other problems the task group identified include:

- Getting longitudinal, community-based studies funded;
- There are not enough good measurements of the efficacy of exercise;
- The secondary conditions that affect exercise among the disabled community, including pain, fatigue, depression, and social isolation;
- A deficit in existing technology to effectively monitor various exercise functions;
- Existing care of chronic diseases, which are a big expense in the health care system and a top cause of death;
- A knowledge gap between scientists and the public, and desensitization of the public to recommendations due to inconclusive evidence that leads to recommendation changes; and
- The existing government recommendations for daily activity are difficult for many Americans to interpret (What is "moderate intensity" or "brisk walking"?).

Future Challenges

The group agreed that further research is needed in the following areas:

- Exploratory development of innovative technology for exercise;
- Measurement of time-duration-intensity or dose-response;
- Measurement of relationship between environment and physical activity;
- Research on novel ways to use technology to create incentives, compliance, motivation;
- Research on how to provide continuity in exercise prescription and

compliance between medical, private sector, and community advocacy organizations; and
• Implementation and tracking of new guidelines from the Department of Health and Human Services.

Some research questions that were developed:

1. Dosage of exercise: Investigate relationship between low levels of activity with health measures.
2. How short can the exercise dose be and remain effective? Optimize exercise dose in order to incorporate effectively into everyday activity.
3. Which outcomes do you pursue? Calories, lean or fat, body weight mass, absolute weight, weight maintenance, cardiovascular fitness, blood pressure, heart rate, flexibility, muscle strength, pain, feelings of wellness?
4. Use technology to motivate people to transition to higher levels or stages of change with respect to physical activity.
5. Enable some progress in metrics toward improving activity level.
6. Create technology that correlates health measures with customizable reward.
7. Education programs: teaching children how to exercise as adults (study of group activity vs. personal activity).
8. Sensors with feedback: clothing, wrist watch, breathing rate, pedometers, schedule with reminders, report to user and/or doctor.
9. Biosensors: muscle tonicity, biochemical or gene response with exercise, cellular response to exercise, blood flow, electrophysiological measure, heat/energy, calories burned, wattage, fat loss, improve insulin sensitivity.
10. How does exercise affect cellular/molecular measures, mechanosensitivity, and stretch-activated channels?
11. Magic drug to motivate people to exercise? Can we identify molecules that motivate people to comply with exercise programs? Improve insulin sensitivity, effect of Viagra on skeletal muscle performance.
12. Activity monitoring in real time: sensitive quantification of activity, prevention of decline, feedback to health care provider, therapists, researchers, and users, and
13. Endocrinology and neuroendocrinology of aging, exercise, and cognition: effect of growth factors, anabolic steroids, hormone supplementation.

The Relationship Between Demographic Mortality Rates, Aging, and Functional Human Healthspan

TASK GROUP DESCRIPTION

Background

Does the demographic plateauing of mortality in late life show us that functional aging decelerates or alters in rate during late life? In the 1990s it was established that age-specific mortality rates plateau at late ages in both humans and well-established experimental organisms, such as *Drosophila*. What remains at issue is whether this constitutes a cessation or slowing of aging itself.

The task group is charged with investigating methods of determining whether mortality rate plateaus constitute a slowing of aging as such, or whether some type of artifact or compositional change is responsible for this novel, important demographic phenomenon. Such methods might employ human medical data, animal model studies, or a combination of the two.

Initial Challenges to Consider

Some standard data on human disability and functional status do not show plateaus, apparently, but this may be in part due to small sample sizes at older ages and to the nature of the measures we use. The most commonly used measures of disability and functional status are activities of daily living (ADL) and instrumental activities of daily living (IADL), and these may saturate at older ages. This is a problem that requires methodological investigation. On the other hand, some data on age-specific incidence of cancers, cardiovascular diseases, and chronic infection show late-life plateaus. Why is there a disparity between this type of health measure and those based on disability?

- To investigate human pathophysiology in late life, do we need functional measures of aggregate aging, for example, performance measures such as walking speed, grip strength, sensory function or cognitive performance, or measures of physiological function? Would other characteristics predictive of mortality also be useful?
- What medical databases could and should be mined for extant information on the time course of human pathophysiology in specific organ systems and tissues, such as heart, brain, liver, and kidney?
- Which noninvasive (e.g., blood, ultrasound) technologies would be most useful in monitoring the physiological transitions that might occur as individual patients undergo the transition from the demographic aging phase to the late-life phase, and age-specific mortality rates plateau?
- The ubiquity of late-life plateaus among humans and well-studied model systems, together with the difficulty of performing experiments on human cohorts, suggests the value of studying the physiology of late life among model animals. The task group is asked to suggest possible model species experimental projects that would explore the question of whether functional aging ceases in late adult life.
- It has recently been established that *Drosophila* age-specific fecundity shows a late-life plateau analogous to that of age-specific mortality rates. Of course, in humans it is difficult to study reproductive effort because of extensive parental care. However, are there useful avenues for studying reproductive pathophysiology in humans during late life.
- Recent work on the evolution of aging in populations with intergenerational transfers may offer useful perspectives on late life in humans. How should evolutionary demographic theory be developed so as to best study the evolution of human late life?
- Do the oldest-old pose a unique environmental design challenge, and how might this challenge be met?

Initial References

Lee, R. D. 2003. Rethinking the evolutionary theory of aging: Fertility, mortality, and intergenerational transfers. Proceedings of the National Academy of Sciences U.S.A. 100(16):9637-9642.

Mueller, L. D., C. L. Rauser, and M. R. Rose. 2007. An evolutionary heterogeneity model of late-life fecundity in *Drosophila*. Biogerontology 8(2):147-161.

Rose, M. R., C. L. Rauser, and L. D. Mueller. 2005. Late life: A new frontier for physiology. Physiological and Biochemical Zoology 78(6):869-878.

Vaupel, J. W., J. R. Carey, K. Christensen, T. E. Johnson, A. I. Yashin, N. V. Holm, I. A. Iachine, V. Kannisto, A. A. Khazaeli, P. Liedo, V. D. Longo, Y. Zeng, K. G. Manton, and J. W. Curtsinger. 1998. Biodemographic trajectories of longevity. Science 280:855-860.

Task Group Members

- X. Edward Guo, Columbia University
- Mary Haan, University of Michigan
- Scott Hofer, Oregon State University
- Bruce Kristal, Brigham and Women's University
- Vikram Kumar, Brigham and Women's Hospital
- Kenneth Manton, Duke University
- Laurence Mueller, University of California, Irvine
- Steven Orzack, Fresh Pond Research Institute
- Judith A. Salerno, National Institute on Aging
- Anatoli Yashin, Duke University
- Ken Wachter, University of California, Berkeley
- Andrea Anderson, New York University

TASK GROUP SUMMARY

By Andrea Anderson, Graduate Science Journalism Student, New York University

The Gompertz curve, named for British mathematician Benjamin Gompertz, predicts that after a certain age, mortality within a population begins increasing exponentially with advancing age. At first blush human mortality rates fit on this Gompertz's law of mortality, making it a common tool for insurance companies and others.

Since the 1990s, though, there has been increasing evidence from large populations that Gompertzian mortality rates may not hold in later years. On the contrary, when scientists evaluate very large cohorts of different species, including medfly,[1] fruit fly,[2] and—arguably—humans, mortality rate levels off after a Gompertzian period. This means mortality does not accelerate indefinitely with age, running counter to the notion that animals senesce unstoppably after a certain point in life. In contrast, mortality may run out of steam in a sense, still claiming individuals within a population but not a higher percentage of them with each passing year.

By recognizing that there are Gompertzian and non-Gompertzian phases of aging, this task group proposed that there may be something unique about late-life biology that makes it possible to relate aging to both mortality and healthspan in new ways. Namely, the group emphasized the importance of understanding what happens to health (specifically impairment and disability) when mortality plateaus. Within this framework they also highlighted the need to find ways of assessing impairment and disability and to better use the measures that are already available.

Charge to Task Group

This task group was charged with devising a plan to address and understand the late-life mortality plateau within the context of human aging and healthspan. This problem raised an interesting set of questions about whether the demographic data represent genuine underlying physiological differences or some sort of statistical artifact. If it is a real effect, does the population's disease load decrease at these late ages along with mortality or are people living with prolonged illness? In short, what can the oldest-old tell us about the process of aging itself?

Strategy

When the task group convened on Wednesday afternoon, it didn't take long for the lively discussion and heated debate to begin. The group initially evaluated the general concept of mortality rate plateaus as well as several potential models for understanding mortality in general. In the process of deliberating over the proposed late-life plateau in mortality rates and what it could represent, the group also discussed a number of different angles from which to approach the problem: at an individual or population level or from a biological, societal, or financial perspective.

Because the task group was charged with trying to understand what happens during the mortality rate plateau, the group spent a long time earnestly discussing whether disability and disease also plateau. This is significant, because if disability increases exponentially while mortality rates plateau, it could mean that individuals are surviving with long-term debilitating illnesses. On the other hand, if disability or disease plateaus, there may be a fundamental difference in the very old that could reveal secrets about aging in general. The state of those in late life is also critical to understanding the question of healthspan and to recognizing the societal and financial implications of late-life mortality plateaus in human populations.

In this context the group thought it was important to draw a clear distinction between disability and impairment. For the purposes of this discussion, the group classified disability as a societal and functional problem, based on a person's circumstances. To some extent disability depends on the technology and assistance available to the individual. Impairment, on the other hand, is related to the loss of a specific function or set of functions that can be measured objectively (e.g., using cognitive or perception tests).

The group also bandied about ideas about just how big a part the underlying biology plays in the late-life mortality plateau. Published data showing a late plateau in age-specific fecundity in large fruit fly populations suggests there are fundamental biological differences in the latest stages of life. As well, one group member noted that the mortality rate plateau was initially interpreted as evidence for plasticity in lifespan. There was a great deal of discussion within the group about this interpretation and how we should define plasticity, if at all. For example, one option was that the mortality plateau might represent more permissive genetic heritage than previously imagined.

The group also discussed whether it is possible to improve people's quality of life at every age and, if so, how function and quality of life should be measured. Some suggestions included everything from technology that assessed voice and motion to cognitive tests to molecular markers such as cytokines. While the group did not reach a conclusion about the most effective tests or how frequently sampling should be, the general consensus was that it would be advantageous to sample as often and as completely as feasible.

One group member noted that in *Drosophila* even with the fecundity plateau, female fecundity decreases rapidly a few days before death, suggesting there may be markers in other species that indicate healthspan decline. If so, there may well be opportunities to manipulate the projected time of death and to decrease the amount of time before death spent with degraded health.

On Thursday morning the task group presented their preliminary report to the other meeting attendees. But the feedback was mixed. Some seemed unconvinced about the demographic evidence for a mortality plateau. Gerontologists pointed out that their oldest-old patients may survive longer, but they continue becoming frailer with time. This countered the task group's assertions that aging itself—including disease load—may be curtailed in late life. Others suggested that selective survival explained the plateau in impairment and declining risk factors observed in some studies.

It was back to the meeting room for more diligent debate. For instance, at least one group member suggested that it might be rash to draw analogies between fruit fly and human plateaus, given less convincing evidence for mortality plateaus in nonhuman mammalian systems such as mice. In the end, though, the group concluded that late-life mortality plateaus were worth exploring further—and in their final presentation the group explained their vision of how to tackle the problem.

Future Challenges and Recommendations

The group concluded that the demographic data suggested potential qualitative differences in late-life biology. For example, not just mortality but also disease load might plateau or even decline in later life. The question is whether these phenomena are related and, if so, how.

To adequately evaluate this, the group recommended creating adequate measures for disability and impairment in late life as well as exploiting the appropriate measures that already exist. For individual and population measures already available, this includes determining the precision, bias, information, and applicability. The group also noted that several kinds of data, assessed on multiple scales, would likely be required to get to the bottom of the mortality plateau conundrum.

Once appropriate measures are available, it should be possible to evaluate the age-associated changes both across populations and within individuals in late life. This could be useful in addressing the broader questions of mortality and healthspan. For instance, better measures of impairment and disease might reveal distinct groups and vulnerabilities within these groups. A variety of noninvasive technologies were proposed—from imaging techniques to cognitive performance tests to the "omics" (e.g., metabolomics, proteomics, and genomics).

Throughout the meeting the group struggled to reconcile the demographic data with the broader concepts of healthspan and aging. It wasn't until they drew up a conceptual model that the problem started to coalesce. The model represents a mathematical interconnection between aging, mortality rate, and healthspan that can theoretically accommodate various kinds of data. For example, age and healthspan can be related to each other based on the absence of disability or impairment, but also other measures of health.

Since mortality is not always a function of age, the process of aging itself could theoretically be broken down into a Gompertz phase and a non-Gompertz phase. This could also help to distinguish between total life expectancy and active life expectancy. The group narrowed its focus to look at the example of a very healthy 95-year-old. They speculated that if it were possible to have perfect knowledge about the individual, it might also be possible to understand the relationship of age with healthspan and mortality.

For instance, if death and health declines are completely stochastic, it might not be possible to predict health declines or death. On the other hand, one mechanism might start an inevitable decline in health ending in death, making it relatively simple to predict both decline and death. Finally, death might only be predictable after an initial health decline. In other words, after health starts deteriorating, it could allow other mechanisms that lead to death.

Similarly, by understanding the oldest-old in our population using centenarian studies we might gain a better understanding of the biology, environmental exposures, and/or lifestyles that are incompatible with making it to the ripe old age of 100. As this task group put it, studying centenarians might "tell us where the mines are buried."

The group also recommended using numerous animal models to try to gauge the physiological markers of healthspan. This approach may hold a great deal of promise in the short term, especially in *Drosophila*, where it is reportedly possible to not only observe but also manipulate the mortality plateau. As a result, the statistical methods may be easier to assess in these model systems. Animal models could also be the key to understanding what is biologically possible.

Finally, the group noted that from a public health perspective, the late-life plateau may have minimal relevance if our society cannot adequately deal with known causes of early death, such as smoking, obesity, malnutrition, and related diseases. This is because much of the impairment and disease burden carried by the elderly results from long-term, chronic exposures that individuals experience earlier in their lives, underscoring the importance of primary prevention in early and midlife for avoiding impairment and disability. Even so, although some continue debating it, the mortality rate plateau presents the intriguing possibility that improving health is possible at any age. And we are just starting to understand just how malleable lifespan—and healthspan—might be.

NOTES

1. J. R. Carey, P. Liedo, D. Orozco, and J. W. Vaupel. Slowing of mortality rates of older ages in large medfly cohorts. Science 258(1992):457-461.

2. J. W. Curtsinger, H. H. Fukui, D. R. Townsend, and J. W. Vaupel. Demography of genotypes: Failure of the limited life span paradigm in *Drosophila melanogaster.* Science 258(1992):461-463.

Changes in Social Contexts to Enhance Functional Status of the Elderly

TASK GROUP DESCRIPTION

Background

Longevity is the largely unexpected consequence of improvements to general living conditions that came from applications of science and technology. Bob Fogel and Dora Costa (1997) coined the term "technophysio evolution" to refer to improvements in biological functioning that are the direct consequence of such advances. They point out that agricultural technologies that were developed mostly in the early 20th century vastly improved the quality and sustainability of the food supply. Subsequent improvements in nutrition were so dramatic that average body size increased by 50 percent and life expectancy doubled. The working capacity of vital organs also greatly improved.

Yet across the same time period there have been relatively few systematic applications of technology and science, including social science, aimed at improving subjective well-being across eight, nine, and even ten decades of life. Indeed, most technological advances currently on the horizon are designed to compensate for age-related impairments like cardiac disease and cognitive decline rather than to improve subjective well-being.

There is good reason to expect that improvements in subjective well-being would not only improve quality of life but would likely reap significant gains in physical functioning as well. In scores of studies subjective well-being has been linked to physical and functional abilities. Indeed, many (e.g., Sapolsky, 2004) have argued that chronic stress in modern life contributes to reduced psychological and physical well-being. Blackburn and colleagues have documented shortened telomeres in caregivers of chronically ill relatives. And the converse: social support has been found to be strongly associated with health and productivity (Berkman, 2000). There is some evidence that positive social relationships may protect against dementia (Fratiglioni, 2000). Happier people appear to live longer and healthier lives (Levy et al., 2000). One finding in particular is clear: high levels of education strongly predict functional health. In order to ensure that added years are satisfying, healthy, and meaningful, it is important to consider ways that scientific and technological advances can contribute to lifelong learning and socioemotional functioning.

At a macro level, too, the attention of economists, demographers, and sociologists is needed. Societies have not adjusted to increased life expectancy. The social norms that guide individuals through life have changed little across the years that life expectancy increased, a phenomenon Riley referred to as structural lag (Riley et al., 1994). This mismatch likely underutilizes the citizenry of the nation. Federal entitlement programs have changed little since their inception despite increases in average life expectancy. Regulations surrounding Social Security have remained largely unchanged, implicitly presuming that people should work for 40 years, and then retire for decades. Medicare reimburses hospital bills but does not reimburse many services that would allow people with disabilities to live at home.

Initial Challenges to Consider

- What is education? Highly educated people show minimal age-related declines in functional status whereas people with less than high school education show steady declines beginning in early adulthood (House et al., 1990). Income contributes to these outcomes but education appears key. What is education? Years spent in the classroom are obviously a gross indicator of education. It is important to gain a better understanding of the cognitive, neural, and behavioral mechanisms that account for improved functioning associated with more years of education.

- How can work/life changes improve subjective healthspan? Technology may offer ways to provide effective continued education throughout life. This will be particularly important for work performance in the future. The speed of transfer of new technologies from discovery to the public is increasing, demanding continual new learning, an area known to decline with age.

- With increases in life expectancy, societies will need aging people to maintain engagement in workplaces, neighborhoods, and families. Can science and technology strengthen families? Families are morphing from horizontal shapes with many siblings, cousins, aunts, and uncles to more vertical family forms. Can technologies strengthen family relations across generations and geographical distances? Cell phones and e-mail changed culture. What are the limits of communications technology to subjective well-being?
- Social functioning is essential to well-being. How can technological and biological advances improve sensation and perception such that social interactions are facilitated into advanced years (e.g., hearing aids, visual aids)?
- How can social programs, such as social insurance programs, be modified to optimize human capital? Education again becomes key. If a shift in responsibility to patient is to work, science education must be improved.

Initial References

Berkman, L. F. 2000. Social support, social networks, social cohesion and health. Social Work Healthcare 31(2):3-14.

Epel, E. S., E. H. Blackburn, J. Lin, F. S. Dhabhar, N. E. Adler, J. D. Morrow, and R. M. Cawthon. 2004. Accelerated telomere shortening in response to life stress. Proceedings of the National Academy of Sciences U.S.A. 101(49):17312–17315.

Fogel, R. W., and D. L. Costa. 1997. A theory of technophysio evolution, with some implications for forecasting population, health care costs, and pension costs. Demography 34(1):49-66.

Fratiglioni, L. 2000. Influence of social network on occurrence of dementia: A community-based longitudinal study. Lancet 355(9212):1315-1319.

Glass, T. A., T. E. Seeman, A. R. Herzog, R. Kahn, and L. F. Berkman. 2000. Change in productive activity in late adulthood: MacArthur studies of successful aging. Journals of Gerontology B—Psychological Sciences and Social Sciences 50B(2):S65-S76.

House, J. S., R. C. Kessler, A. R. Herzog, R. P. Mero, A. M. Kinney, and M. J. Breslow. 1990. Age, socioeconomic status, and health. The Milbank Quarterly 68(3):383-411.

Levy, B. R., M. D. Slade, S. R. Kunkel, and S. V. Kasl. 2000. Longevity increased by positive self perceptions of aging. Journal of Personality and Social Psychology 83(2):261-270.

Riley, M. W., R. L. Kahn, and A. Foner (eds.). 1994. *Age and Structural Lag: Society's Failure to Provide Meaningful Opportunities in Work, Family, and Leisure.* New York: John Wiley.

Sapolsky, R. M. 2004. Organism stress and telomeric aging: An unexpected connection. Proceedings of the National Academy of Sciences U.S.A. 101(50):17323-17324

Task Group Members

- Richard Allman, University of Alabama at Birmingham
- Eileen Crimmins, University of Southern California
- Corey Keyes, Emory University
- Steven Kou, Columbia University
- Kenneth Langa, University of Michigan
- Duncan Moore, University of Rochester
- Greg O'Neill, National Academy on an Aging Society
- Sara Peckham, Wellness Coordinator
- Teresa Seeman, University of California, Los Angeles
- Rachel VanCott, Massachusetts Institute of Technology

TASK GROUP SUMMARY

By Rachel VanCott, Graduate Student in Science Writing, Massachusetts Institute of Technology

Mr. McGuire: I want to say one word to you. Just one word.
Benjamin: Yes, sir.
Mr. McGuire: Are you listening?
Benjamin: Yes, I am.
Mr. McGuire: Plastics.
Benjamin: Just how do you mean that, sir?
From *The Graduate* (1967)

An Aging Society Requires Social Plasticity

After three days of discussion, this task group started their presentation with a joke from the movie *The Graduate* and a nod to ideas introduced in the keynote address just days before.

People are living longer, said group spokesperson Kenneth Langa, but they're living those longer lives according to an outdated set of social norms. Today's social structures support a lifecourse that leads to stagnation: Education is confined to the early years of a person's life, his adult years are dominated by repetitive work, and leisure finds a place only at the end of life, during retirement. The task group had been charged with considering how changing social contexts could improve the functional status and subjective health of the elderly. To do that the group looked beyond retirement and across the lifespan.

They considered how increasing options for education, work, and leisure throughout life might extend health. Through their discussions the group decided that the key to a society that keeps individuals engaged and learning throughout life is a restructuring of the lifecourse: a move toward social plasticity.

The main focus of the group's work was figuring out how to break down barriers between the life stages and allow opportunities for education, work, and leisure to permeate the lifespan, summarized Langa, who is a professor of general medicine and of the Institute for Social Research at the University of Michigan.

Day 1: Restructuring Education

The task group spent much of their first meeting discussing education, and the first question of the presentation voiced a knowledge gap that the group identified early in the conference. "What is it about education early in life and possibly later in life that allows people to maintain their minds, brains, and bodies optimally?" asked Langa, "Is it the formal process of education? Is it about the social interactions that happen in the education process? How do we optimize the active ingredients?"

Several group members noted that it might not be education itself that results in high levels of functioning later in life. Those who succeed in higher education might have other characteristics associated with better late-life health. For example, the drive that leads people to seek and complete higher education may also encourage other behaviors, like disciplined exercise habits, which lead to better health.

While the group members weren't sure how many factors contribute to the relationship between education and high functioning in later life, several of them argued that the relationship has been shown to exist. There is plenty of research that shows that education itself has a significant impact on the brain, said Langa. "A more highly educated person can sustain more damage to the brain . . . as circuits start to fail, the brain [of an educated individual] can reroute to use other circuits."

So even without knowing how and if the formal education process is directly responsible for functional enhancement, public policy decisions that make higher education available to people of any age might be a good place to start, suggested Greg O'Neill, group leader and director of the National Academy on an Aging Society. The group discussed several factors that make accessing education at older ages difficult, including transportation issues and the social stigma of attending school as an adult. The group also identified the rising cost of higher education as a potential barrier that public policy decisions could impact.

"The government is giving funding for young people who want to go to school but not for older people who want educational experience," said Steven Kuo, an associate professor of industrial engineering and operations research at Columbia University.

Brian Hofland, program director of the international aging team of the Atlantic Philanthropies Incorporated, spoke up, explaining that programs do exist to serve that purpose. For example, he said, the Council for Adult and Experiential Learning, a nonprofit organization based in Chicago, Illinois, has a program called Lifelong Learning Accounts (LiLA), which allows employers to match funds that the employee could use for higher education or training purposes.

Teresa Seeman, a professor of geriatrics at the University of California, Los Angeles, argued that older adults seeking higher education still have less financial aid options than young adults. But educational opportunities don't occur exclusively in the classroom. Seeman mentioned the existence of a number of small programs that offer the chance for older adults to gain various new skills or knowledge. O'Neil also mentioned a learning opportunity far from the classroom—the organization Encore! sends volunteers who have previously worked with the Peace Corps out on short-term assignments where they can use their skills and experience and in turn continue to learn more about the world and about themselves.

However, as promising as the opportunities for alternative types of learning sound, Seeman said, they haven't been evaluated with regard to their impact on a person's physical or mental health.

Seeman's comment reflected a concern that would surface again and again over the three days of group discussion. Many organizations in the private and public sectors are proposing and producing ways to keep older adults learning longer, but if the results of these programs aren't objectively evaluated, scientists and policy makers will be unable to answer the pressing questions surrounding this issue. Is formal education that happens later in life as beneficial as education obtained as a young adult? Do alternative learning opportunities that teach skills and self-knowledge provide the same cognitive benefits as formal education?

Day 2: Restructuring Work and Leisure

On the second day the group returned to an issue they had touched on during the first day: work and retirement.

"We know the modal age of retirement was 62, which is starting to creep up." said Eileen Crimmins, who is a professor of gerontology at the Davis School of Gerontology at the University of Southern California and director of the USC/UCLA Center on Biodemography and Population Health. "People want to retire at some point," she said, "but they like the idea of having jobs and flextime. We know what people want to do and we kind of know what people have to do."

What people have to do, the group agreed, is continue working later in life. That might mean later retirement or second careers or part-time work while in retirement. To encourage continued work older adults would need more options, including the chance for midcareer retraining and the opportunity to have different types of intellectually or socially meaningful part-time jobs as a second career.

Opportunities like working with Experience Corps, a program funded by AmeriCorps that trains older adults to act as tutors for struggling city-school students, might provide the type of opportunity that could keep the elderly socially and intellectually engaged while they earn a small stipend. But Experience Corps, like other programs, hasn't been evaluated for the health impact it has on the adults who tutor.

In addition to a lack of appealing job options, transportation can be a major barrier to continued work, particularly in the case of older adults who may be unable to drive or rely on public transportation to take them to work and social gatherings, especially in extreme climates. The group discussed technology-fueled alternative work ideas like microjobs—jobs that can be done in 10 to 15 minutes from a home computer—or work at call centers, or programs in which call-center calls can be redirected to home phones.

Outside of offering greater opportunities for work in the home, technology might also enhance the social life of the elderly by allowing them to stay in touch with family and friends, or allow them to maintain autonomy longer through use of smart house systems that could monitor their movements and alert family if their movement patterns change. Alternatively, a new type of intergenerational living group, or urban housing center, for the elderly might help older adults remain socially engaged even if their families live far away.

But just like the examples that the group had discussed previously, technological innovations and alternative living arrangements have not been assessed for their potential as tools that help the elderly maintain their subjective well-being.

Recommendations

In the final presentation Langa concluded with the group's recommendation. "We need to gather data about what older adults actually want in terms of new opportunities for ongoing engagement, productivity, and learning into older age," he said. "What are the preferences of people, and are there differences in preference over various social groups? Are there significant differences across class, across socioeconomic status?"

Langa also introduced the concern that had become a theme in this task group's discussion. "We don't think we have a great sense of what the currently existing programs are," he said. "We'd like to put that information together and gather what's out there in terms of current programs . . . evaluate these programs and their consequences in terms of health and healthspan." The group also suggested that a prize, designed to award programs that successfully improve the functional status of the elderly, might spur further research.

Overall, the group suggested that our social structures need greater plasticity so that they can better meet the needs of both our aging society as a whole and the needs of individuals as they grow and change. Organizations that are already working toward this kind of plasticity must be recognized and encouraged if we want to see an increased human healthspan and push the limits of human health.

Develop Technological Interventions to Overcome Barriers to Independence and Community Participation

TASK GROUP DESCRIPTION

Background

Technology has and will continue to affect people in different settings with different levels of functioning. The setting may be at home, providing personal support and help for daily living. It could be a neighborhood, where the systems help a person to engage in community activities. Or it could be more societal, where a person commutes to work and contributes to society through employment. In each setting, technology may provide different forms of functionality: enhancing dexterity and mobility, helping with some home chores, supporting aspects of memory, coaching through particular job functions, and helping to drive vehicles, for example. In other words, technology may touch almost all aspects of human living.

Initial Challenges to Consider

- Identify technological interventions needed for people to continue to live active lifestyles and to support healthy aging with a disability.
- What are the barriers (social, cultural, socioeconomic status, policy, clinical) to using current and future technologies?
- Where can technology be applied most effectively for education, monitoring, health promotion, and to enhance quality of life?
- How do we effectively build teams to create new technologies

to incorporate broad scientific and technological expertise and include consumer/caregiver participation?

- What are the areas in which technology is most needed: personal assistance, caregiver assistance, mobility, cognition, perception, awareness, sensation, vocation?
- How can universal design principles be applied to increase technology acceptance and diffusion?
- How can advances in robotics, artificial intelligence, rapid prototyping, machine learning, materials science, computing, activity monitoring, and modeling, etc. be adopted to benefit older people and people with disabilities?
- Develop tools/measures/metrics to assess the impact of technology on activity, quality of life, cost, community participation, and health (beyond morbidity and mortality).
- What are the critical biological and biocompatibility issues that need to be overcome?
- Develop technological innovations that promote independence, community participation, and healthful living.

Initial References

Chavez, E., M. L. Boninger, R. Cooper, S. G. Fitzgerald, D. Gray, and R. A. Cooper. 2004. Application of a participation system to assess the influence of assistive technology on the lives of people with spinal cord injury. Archives of Physical Medicine and Rehabilitation 85(11):1854-1858.

Collins, D. M., S. G. Fitzgerald, N. Sachs-Ericsson, M. Scherer. R. A. Cooper, and M. L. Boninger. 2006. Psychosocial well-being and community participation of service dog partners. Disability and Rehabilitation: Assistive Technology 1(1-2):41-48.

Cooper, R. A. 2004. Bioengineering and spinal cord injury: A perspective on the state of the science. Journal of Spinal Cord Medicine 27(4):351-364.

Cooper, R. A., R. Cooper, M. Tolerico, S. F. Guo, D. Ding, and J. Pearlman. 2006. Advances in electric powered wheelchairs. Topics in Spinal Cord Injury Rehabilitation 11(4):15-29.

Iezzoni, L., E. McCarthy, R. Davis, and H. Siebens. 2001. Mobility difficulties are not only a problem of old age. Journal of General Internal Medicine 16(4):235-243.

Leuthardt, E. C., G. Schalk, D. Moran, and J. G. Ojemann. 2006. The emerging world of motor neuroprosthetics: A neurosurgical perspective. Neurosurgery 58:1-14.

Odenheimer, G. 2006. Driver safety in older adults: The physician's role in assessing driving skills of older patients. Geriatrics 61(10):14-21.

Parasuraman, A. 2000. Technology readiness index (TRI): A multiple-item scale to measure readiness to embrace new technologies. Journal of Service Research 2(5):307-320.

Pearlman, J., R. A. Cooper, E. Zipfel, R. Cooper, and M. McCartney. 2006. Towards the development of an effective technology transfer model of wheelchairs to developing countries. Disability and Rehabilitation: Assistive Technology 1(1-2):103-110.

Talbot, L. A., J. M. Gaines, T. N. Huynh, and E. J. Metter. 2003. A home-based pedometer-driven walking program to increase physical activity in older adults with osteoarthritis of the knee: A preliminary study. Journal of the American Geriatric Society 51(3):387-392.

Trudel, T. M., W. W. Tryon, and C. M. Purdum. 1998. Awareness of disability and long-term outcome after traumatic brain injury. Rehabilitation Psychology 43(4):267-281.

Winters, J. M. 2006. Future possibilities for interface technologies that enhance universal access to health care devices and services. In *Medical Instrumentation: Accessibility and Usability Considerations,* 1st ed., eds. J. M. Winters and M. F. Story, pp. 321-339. Boca Raton: Taylor & Francis, CRC Press.

TASK GROUP DESCRIPTION—GROUP A

Rewritten by Noah Barron, Graduate Journalism Student, University of Southern California

(As stated in the preface, "Some groups decided to refine or redefine their problems based on their experiences." This Task Group Description was rewritten by Noah Barron to reflect Group A's decision to redefine the challenge at hand.)

Background

Imagine for a moment the senior citizen of the future. She lives in a smart home that's tailored to her every need, engineered to keep her healthy, active, and independent. Sensors tuned to her specific biometrics read her mood and whereabouts. The flooring is ready to dissolve into a cushioning gel in case she takes a potentially hip-shattering fall. The appliances wirelessly talk to one another to cook her meals and automatically order more milk when she's running low.

All of these technological marvels serve a common purpose: to keep our senior independent. A residence like the one above would allow her to live on her own longer and delay institutionalization, perhaps indefinitely. Once a person is committed to nursing care, health and mental aptitude tend to rapidly decay. Staying out of that system and aging in place seems to be the key to living longer, healthier lives, explain senescence experts who study the mechanics of advanced-age living.

But what exactly does independence mean? The task group defined the necessities of aging in place along the following lines: assistive technologies for the augmentation of healthy aging, for cognition and mobility, for the activities of daily life (basic and instrumental), for chronic disease amelioration, to alleviate social isolation, to lessen the burdens of caregivers and family, and to increase personalization of care.

Any technology of the future should do all of the above in a way that doesn't simply replace human care with mechanized care; rather, it should enhance the ability of a few humans to care for many while maintaining the strong social bonds that keep seniors healthy and mentally alert.

The task group began with specifics. What ends a senior's independence? Surprisingly, the tipping catalyst for institutionalization isn't senile dementia or Alzheimer's. It's loss of toileting independence. In America, where most senior care is provided by family members, worn-out offspring often draw the line at cleaning up after their own parents.

A 2003 study by the School of Public Health in Tampere, Finland, found that urge incontinence was the most significant predictor of coming institutionalization. When seniors can no longer make it to the bathroom in time, get on and off of the toilet by themselves, and maintain hygiene, their caregivers and relatives ship them off to the home, where the downhill slide of mental and physical deterioration accelerates rapidly.

The first suggestion—perhaps more as a mental exercise than a literal technology—was to create a toilet that helps seniors on and off, reducing fall risk and lifting/hygiene work by caregivers. Using the toileting problem as a jumping-off point, what sorts of flexible solutions to independence problems can future technology provide?

Initial Challenges to Consider

First, the group set out potential barriers to creating new assistive products for the senior citizen's smart home. Is the technology ready? If not, how far down the line is it? What will it cost? How will it be tested? How will its efficacy be measured? Will it end up unused in the closet? Is it dangerous? How will it communicate with other technologies? If it malfunctions, who will fix it? The answers to questions like this will determine whether a technology is actually viable.

One way of parsing the problem is to think about it in terms of replacement technologies versus disruptive technologies. A replacement technology would be a better wheelchair that takes the place of an existing wheelchair technology, whereas a disruptive technology would be one that fills a niche that was previously vacant and rapidly diffuses throughout society (e.g., iPod, BlackBerry, GPS). The group chose to focus on disruptive technologies because they offer the greatest opportunity for innovation as well as more complex responsibilities as the conceptual level for creating something that's a genuinely beneficial idea (e.g., health benefits) weighed against ethical and privacy concerns.

Additional References

Collis, G. M., and J. McNicholas. 1998. A theoretical basis for health benefits of pet ownership: Attachment versus psychological support. In *Companion Animals in Human Health*, eds. C. C. Wilson and D. C. Turner, pp. 105-122.

Hidler, J., and G. Hornby. nd. Gait restoration in hemiparetic stroke patients using goal-directed, robotic-assisted treadmill training. NIDRR project overview. http://www.smpp.northwestern.edu/MARS/Project2descII.htm, accessed Feb. 14, 2008.

Krebs, H. I., N. Hogan, B. T. Volpe, M. L. Aisen, and C. Diels. 1999. Overview of clinical trials with MIT-MANUS: A robot-aided neuro-rehabilitation facility. Technology and Health Care 7(6):419-423.

MacDorman, K., and H. Ishiguro. 2006. The uncanny advantage of using androids in cognitive and social science research. Interaction Studies 7(3):297-337.

Matarić, M., J. Eriksson, D. Feil-Seifer, and C. Winstein. 2007. Socially assistive robotics for post-stroke rehabilitation. International Journal of NeuroEngineering and Rehabilitation 4(5).

Mori, M. 1970. The uncanny valley. Energy 7(4):33-35.

Nuotio, M., L. Teuvo, J. Tammela, T. Luukkala, and M. Jylha. 2003. Predictors of institutionalization in an older population during a 13-year period: The effect of urge incontinence. Journals of Gerontology A—Biological and Medical Sciences 58:M756-M762.

Stiehl, W. D., and C. Breazeal. Forthcoming. Affective Touch for Robotic Companions. Presented at First International Conference on Affective Computing and Intelligent Interaction, Beijing, China.

Stiehl, W. D., J. Lieberman, C. Brezeal, R. Cooper, L. Knight, L. Lalla, A. Maymnin, and S. Purchase. 2006. The huggable: A therapeutic robtoic compational for relational, affective touch. Consumer Communications and Networking Conference, CCNC 2006. 3rd IEEE 2(8-10):1290-1291.

Wada, K., T. Shibata, T. Saito, and K. Tanie. 2002. Effects of robot assisted activity for elderly people at day service center and analysis of its factors. Presented at 4th World Congress on Intelligent Control and Automation, Shanghai, China.

Due to the popularity of this topic, two groups explored this subject. Please be sure to explore the other write-up, which immediately follows this one.

TASK GROUP MEMBERS—GROUP A

- Stephen Abramowitch, University of Pittsburgh
- Rory Cooper, University of Pittsburgh
- Clifford Dasco, Methodist Hospital Research Institute
- Cristina Davis, University of California, Davis
- Arun Hampapur, T. J. Watson Research Center
- Maja Matarić, University of Southern California
- Hunter Peckham, Case Western Reserve University
- Jonathan Wanagat, University of Washington
- Mike Weinrich, National Center for Medical Rehabilitation Research
- Noah Barron, University of Southern California

TASK GROUP SUMMARY—GROUP A

Summary written by Noah Barron, Graduate Journalism Student, University of Southern California

One possible disruptive tech solution for the problems of isolation and lack of independence for seniors aging at home would be assistive robotics.

Enter our futuristic live-in nurse robot, the hypothetic centerpiece of tomorrow's smart home. Picture Rosie, the Jetsons' automaton maid, but specially programmed to care for older people.

Instead of struggling to get in and out of bed unassisted, the robot could lift the old person in carefully. Changing sheets? Help up onto the toilet? No problem. Autonomous assistive robots could take the place of human live-in caregivers. And if something were to go truly wrong, such as a fall or a stroke, intelligent robots could give first aid and contact emergency services.

But the value goes deeper than that, say scientists. One of the most deadly and tragic parts of aging is social isolation. Countless studies have shown that the more people and interactions you have in your life in its later years, the longer and healthier you live.

For many, a wide social network of humans is not an option. Intelligent social robots could provide the interaction that old people need to stay sharp.

Currently there are several robotics rehabilitation projects in development that could bring robotics to the forefront of old-age care and rehabilitation. It's a surprising menagerie of robotic healthcare workers. What follows is by no means a comprehensive list of current assistive technologies; rather, it is a cast of potential characters that may populate the smart homes of the future. At this point we can only imagine the possibilities; an overview of what is available today is helpful for that speculative exercise.

The Massachusetts Institute of Technology (MIT) created Manus, a robotic physical therapy machine that rehabilitates stroke victims. Manus instructs them on how to perform hand exercises and assists them in moving their muscles if they cannot do so unaided. However, patient questionnaires from the project indicate that people liked working with Manus but would have preferred guidance from a human therapist.

Over in Switzerland, medical robotics engineers are developing Lokomat, a sort of mechanized lower-body robot suit that helps stroke and semiparalyzed patients relearn to walk by putting their legs and torso through the motions of a natural stride. Again, this technology exists primarily to replace the hard physical labor that rehab therapists do, lifting and manipulating patients' bodies.

University of Southern California social robotics researcher Maja Matarić hopes that robotic integration in the home will be more personal than lifting, washing, and helping. She imagines a world where robots provide social and emotional encouragement as well as physical assistance. But we're not there yet, she said.

For one, Matarić's autonomous robots have a strict no-touch policy because their control software isn't reliable enough to ensure they won't hurt anyone. In other words, they can't be trusted to follow Isaac Asimov's first law of robotics: "A robot may not injure a human being."

"We have a no-touch safety rule because a robot strong enough to get you out of bed could crush you, could kill you," said Matarić.

Matarić and other social robotics engineers are keeping their robots small so they can't injure humans. "Why do you think all those toylike Japanese robots are tiny?" she said. "It's to keep them weak."

Honest researchers admit that robotics and human interaction is still not terribly advanced. "Talk to anyone in robotics when they are not drunk and they will tell you the same thing," Matarić said.

For now, diminutive droids can offer encouragement, challenge patients to rehabilitate themselves, and offer sensory stimuli to folks who would otherwise be shut in without anyone interesting to interact with.

The current goal of social robotics, especially in the context of companion caregivers for the aged or the disabled, is to figure out exactly how to inspire that feeling of empathy in people. Ironically, the less convincingly humanoid the robot, the more likely people are to feel affection for it.

In the words of Cliff Dasco, director of the Abramson Center for the Future of Health, Methodist Hospital Houston, "It's the R2D2 effect—everyone likes him more than C3PO." (In case you slept through *Star Wars*, R2 is the lovable garbage-can droid and C3PO is the annoying humanoid one.)

One theory that was advanced in the 1970s is that there is a point where humanoid robots stop seeming like cute automata imitating people and start seeming like people with something hideously wrong with them.

This point was called "the uncanny valley" by Japanese robotics engineer Masahiro Mori. The "valley" is the massive dip in the chartable empathy response people feel toward androids as they become more lifelike. The more human, the more we like them, until suddenly the robots start really creeping us out. He argued that the tipping point happens when a robot gets so close to humanlike that we stop focusing on its similarities to us and become fixated on its differences.

It tilts its head strangely, speaks with a certain lack of warmth or has no ephemeral sparkle in its eyes. We stop thinking of it as an advanced robot and start to feel like we are chatting with an animated corpse.

Mori thought that the uncanny valley is an evolutionary response that's designed to tell us to steer clear of people with diseases or disabilities that could harm us.

As such, some therapeutic robotics developers have moved away from near-human robot design to avoid falling into the valley, choosing to focus on assistive bots that are furry and cuddly.

MIT's Huggable robotic companion is a furry teddy bear intended for children's therapy. It's well studied that petting a kitten or puppy can lower your heart rate, reduce stress, lift your mood, and rehabilitate your social skills, but pet therapy often isn't available because many patients have allergies, institutional or hospital rules may forbid pets—or simply because a troubled or motor-impaired patient would harm the animal.

Huggable conceals an advanced set of instruments housed in its fuzzy exterior. Beneath the fur and silicone of Huggable, sensate skin sensors analyze the temperature and electric field of the patient who's holding it, as well as how much pressure he or she is putting on the bear—indicators of health and stress.

Cameras in the eyes, microphones in the ears, and an onboard wireless

PC connected to teleoperated controls allow remote therapists to observe how the patient is acting, talking, and petting it and servos in the face and neck give Huggable the ability to gently and silently move its head and make expressions in response. In short, it is nonthreatening, extremely cute, soothing, and can simulate affection toward the patient while relaying valuable diagnostic information to the doctors.

The Road Ahead

Technologies such as those described above are just the beginning. Assistive robotics is simply one avenue among many. The task group set out a rough chart of available assistive technologies for senior independence and plotted where they are now and where each is likely headed in the years to come, as well as obstacles faced and criteria for measuring success (Table 1).

TASK GROUP MEMBERS—GROUP B

- Lazelle Benefield, University of Oklahoma Health Sciences Center
- Leon Esterowitz, National Science Foundation
- Stuart Harshbarger, Johns Hopkins University
- James Kahan, RAND Corporation
- Russell E. Morgan Jr., SPRY Foundation
- Margaret Perkinson, Saint Louis University
- Thomas Zimmerman, IBM Almaden Research
- Jane Liaw, University of California, Santa Cruz

TASK SUMMARY—GROUP B

By Jane Liaw, Graduate Science Writing Student, University of California, Santa Cruz

As we catapult into the 21st century and beyond, it will take a village not only to raise a child but also to keep that child healthy into old age. Our neighborhood is no longer the few blocks or miles we traverse every day between work and home—it is the global society to which we belong, a world whose fabric is woven tighter every day.

That we live in this ever-diminishing world is not news. We are reminded of it constantly, as we read globalization and outsourcing stories. Yet it is not enough to recognize our changing universe; scientists must

TABLE 1 Assistive Technologies for Senior Independence

Technology	Current Status	Near Future
Adaptability aids	Walkers, scooters, motorized wheelchairs, iBot	Automated cars, powered exoskeleton
Socially assistive robots	Demonstration, motivating rehab, exercise, petlike toys	Robotic assessment of patient health
Robotics and sensor technologies	Controlled environments, teleoperation, surgery	No-contact autonomy, teleop caretaking
Biosensors	Smart toilet, miniature mass spectrometer	Ubiquitous computing, mobile diagnostic lab on a chip
Cognitive augmentation	Reminders (as on pill box), cognitive orthotics, text-to-speech	Unknown
Activity monitoring	Motion sensors, cameras, radio freq identification, data extraction and pattern recognition	Lightweight, wearable, real-time behavior recognition, alerting

respond to it, and respond quickly. The scientist's training is usually the antithesis of a global approach—the more scientific education one has, the more likely one is to be a God of Small Things, specializing in one tiny area of knowledge.

While experts are of course necessary, the problems we face in the new world require us to zoom out and see every issue in its entirety. Aging is an especially ripe challenge for a multidisciplinary approach, since it cuts across many fields and affects everyone. Our group was charged with putting our collective minds and disparate backgrounds to a topic that let us dream far and dream big: How could we use technology to overcome barriers to independence for the aging?

Before we could jump into discussions, we first had to define the scope of our question. If we are to talk about how technology might help the disabled, for example, what do "technology" and "disability" mean in this con-

Far Future	Barriers	Metrics
Slim form-factor exoskeleton	Cost, acceptance, safety, miniaturization, power source	Market, the "thrown in the closet" factor, other pernicious outcomes
Multifunction technologies, human contact	Cost, acceptance, the uncanny valley	Market, closet factor, health outcomes, quality of life
Autonomous manipulation, toilet care/hygiene, moving people	Cost, acceptance, the uncanny valley	Market, closet factor, health outcomes, quality of life
Nanotech, cellular sensors, autonomous function	Cost, acceptance, safety, miniaturization, power source	Market, closet factor, pernicious outcomes
Neural implantation	Innovation, lack of basic knowledge, interface issues, ethics	Cognitive performance, quality of life
Long-term trending prediction, identifying health risk predictors	Privacy, cost, acceptance, data interpretation and mining	Health outcomes, market value

text? Are we thinking of technology as sophisticated engineering products like robots that can automate many tasks for the elderly? Or is it more?

"Where does disability exist?" asked Margaret Perkinson, a medical anthropologist and associate professor of occupational science and occupational therapy at St. Louis University. "Disability is an interaction with a particular environment and the challenges that particular environment provides."

Russell Morgan, president of the SPRY Foundation, preferred a broad focus of disability. "In most minds disability is physical, but it could also be economic or cognitive. Low socioeconomic status is a disability, and technology can pull disabled people together."

The group ended up deciding that both disability and technology were flexible concepts best defined as broadly as possible. Technology would not just be robots and complicated machinery but also many tools—even simple, existing tools such as pill grinders that promote easy pill swallowing—or health measures not typically considered technology, such as vaccines.

Next, we asked what constituted a barrier to independence. Again, we decided the word could mean many things. The aging may deal with not only physical barriers such as hearing, vision, or physical dexterity losses but also barriers caused by poverty, lack of political clout, or lack of awareness about options. Often overlooked are societal attitudes that might create barriers, such as ageism, or prejudice against older people. This prejudice leads some seniors to avoid using devices that would advertise their disabilities, such as hearing aids.

We concluded that technology developers should be sensitive to the need for their clients to maintain dignity. If the elderly do not want to let the world know they need hearing aids, then designers should create less conspicuous hearing aids, perhaps by making them look like cell phone earpieces.

Older people are sometimes afraid of new technologies, and this attitude can also be a barrier. They might fear computers, for example, as being too complicated or impersonal.

"A key point is making the products friendly for end users," said Lazelle Benefield, a professor of gerontological nursing at the University of Oklahoma Health Sciences Center.

We kept Benefield's words in mind as we divided the technologies into three horizons—currently available, in development, or still a dream. We allowed ourselves to talk about both the practical and the far-fetched.

The first horizon is full of what our team leader, Thomas Zimmerman, called "low-hanging fruit." Zimmerman, an engineer with IBM Almaden Research Center, categorized these currently available solutions as generally low-tech, and not as widely used as they could be. In this group are the vaccines and pill grinders, as well as sensors to monitor falls or wandering, computer literacy, and so on.

"With any technology there's a real issue of access," said James Kahan, an adjunct behavioral scientist at RAND Corporation.

These first-horizon technologies are often not accessible to or are underutilized by seniors, an indication that experts in different fields should be communicating and sharing knowledge more.

When he looked at our first-horizon list so far, Stuart Harshbarger asked, "Where in horizon 1 is active disease management?" Harshbarger, a system integrator with the National Security Technology Department at

Johns Hopkins University, added automated remote monitoring of blood pressure, weight, and glucose levels to the list.

Zimmerman suggested repurposing existing technology for use by aging populations. Video games, such as Dance Dance Revolution or Wii games, that require players to be physically active could be tweaked so older players could enjoy using them for exercise. The group thought the idea of reappropriating technology was an efficient solution and could be applied in other situations as well.

The second horizon comprises technologies that are right now in trial phase. The challenge with these emerging works is in picking and supporting the winners and then paving the way for their distribution. That path requires jumping through hoops such as streamline testing, Food and Drug Administration approval, marketing, and liability issues. Some suggestions in this horizon are relatively straightforward to implement. Screening tests can be improved—such that colonoscopies are prep-free and comfortable, or diseases can be detected through the breath—and community areas can be designed to promote healthy lifestyles. Leon Esterowitz, program director of the Chemical, Bioengineering, Environmental, and Transport Systems Division at the National Science Foundation, suggested machines to assist patients with balance problems and robots to help the elderly exercise.

Other second-horizon solutions face more complicated journeys to mainstream acceptance; before we can use nanotechnology and stem cells for cartilage rejuvenation, for example, there may first be years of policy debates.

The third horizon of dreamscapes and science fiction lets us discuss high-risk, high-reward ideas without concern for their immediate feasibility. This was the realm of artificial intelligence brain implants to offset cognitive decline, artificial retinas, regenerated limbs and organs, robotic caretakers, and behavior-changing technology. In the last example the technology could be used with young people also, to eliminate addictions, reduce obesity, and increase exercise so they grow into healthy old age.

As we got into the meat of the discussion, it became clear some people were more focused on specifics, some more on big picture generalities. Similarly, some leaned toward taking advantage of low-hanging fruit and some wanted to concentrate on projects that now exist only in the imagination. It took some time on the first day for the dialogue to meet in the middle.

Kahan brought in the concept of the innovation cycle, which then became the basis of our brainstorming. In this cycle, science is only one piece, neither the beginning nor the end of the process. From the science we create products and technology that are then produced for public consumption. For the public to get their hands on the products, they must be aware of and have access to them.

But the innovation process doesn't end with the user. As Kahan reminded us, "We have to consider all six points of the innovation cycle" (Figure 1). An important step that is sometimes neglected in the real world is user feedback that would improve the product so that it better meets consumer needs. And that feedback must be taken into account as scientific research continues.

If feedback is so important in our model, how should it be collected? We segued naturally into a discussion of tools and measures.

There were several possible measures, we concluded. User opinions could be solicited via websites, interviews, studies, and certifications similar to the Good Housekeeping Seal of Approval. Demand in the marketplace has always been a measure of success. Yet market forces need to be balanced with social justice needs, especially when health issues are in question. Government policies would be an avenue to protect those needs.

With our innovation cycle still in mind we envisioned the technology

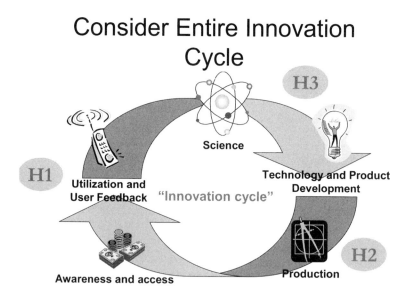

FIGURE 1 The innovation cycle.

creation team as much more than just biological scientists and engineers. An innovation model that puts a premium on user feedback would include advocates for older adults, social scientists who study how people incorporate technology into their lives, ethicists, policy experts, business people, and end-user evaluators, among others. It would be a team that looks at aging solutions from all angles.

On the second day we gathered with group A. Zimmerman and group A's representative took turns presenting our results so far. It appeared that we had taken a different tack from them—they focused mainly on information technology and robotics, while our broader definitions of disability and technology meant we were looking at many other areas. We agreed that the two groups would complement each other in the final presentations.

From our brainstorming and flights of fancy came two concrete recommendations to push barrier-removing technology forward. The first was a Web-based clearinghouse that consolidates information on available technologies for the aging and allows user input, *a la* Amazon.com or Netflix. This not only gives users a voice but also keeps the database current.

The second recommendation was to organize and identify key participants for an international conference on creative ways of harnessing computer-based technology to achieve healthy aging. These technological tools would enable older adults (people 50-75 years of age) to improve their health and avoid diseases. Experts would share knowledge on currently available technology, priorities for future developments, and funding sources. Conference organizers could sponsor competitions for young scientists to develop innovative technologies.

Our group gelled on the second day: though the atmosphere was still respectful, members were more comfortable in healthy debate. We had not expected, with just two days together, to build a miracle robot or even to delve deeply into any aspect of our challenge. We did, however, begin a conversation among scientists from various fields and practice harnessing knowledge from each participant to contribute to a greater whole. We defined a problem together and came to a consensus on approach.

When Zimmerman presented our group's findings on the final day of the conference, he referred to the "age tsunami" coming our way, due to the large group of baby boomers becoming seniors. It is a daunting challenge for researchers and policy makers, but perhaps there is a silver lining—perhaps we will finally expand our knowledge on aging and change perceptions of what aging should mean. In the 21st century we can no longer shunt the elderly to a corner and expect them to quietly fade away. As baby boomers reach their golden years their voices will be heard loud and clear—and we must learn to listen and respond.

Additional Reference

Dethlefs, N., and B. Martin. 2006. Japanese technology policy for aged care. Science and Public Policy 33(1):47-57.

Cellular and Molecular Mechanisms of Biological Aging:
The Roles of Nature, Nurture, and Chance in the Maintenance of Human Healthspan

TASK GROUP DESCRIPTION

Background

The degree to which an individual organism maintains healthspan and lifespan is a function of complex interactions between genetic inheritance (nature), environment, including cultural inheritance (nurture), and stochastic events (luck or chance). This task group will focus upon the role of chance because it is so poorly understood and because it appears to be of major importance in the determination of individual variations in healthspan and lifespan *within* species. The major factor determining variations in healthspan and lifespan *between* species is genetic inheritance. Broader aspects of cellular and molecular mechanisms of biological aging will also be considered, given their importance for understanding the cellular and molecular basis of successful aging. The task force will consider the cellular and molecular basis for nature, nurture, and chance in healthspan and lifespan determination.

On the basis of comparisons between identical and nonidentical twins, geneticists have estimated that genes control no more than about a quarter of the interindividual differences in lifespan (Herskind et al., 1996). Twin studies of very old individuals, however, show substantially greater genetic contributions to healthspan (McClearn et al., 1997; Reed and Dick, 2003). The environment clearly plays an important role in the length and the quality of life. Tobacco smoke, for example, has the potential to impact upon multiple body systems in ways that appear to accelerate the rates at which those systems age (Bernhard et al., 2007).

To document the role of chance events on aging one must rigorously control both the genetic composition of an organism and its environment. This has been done to a remarkable degree in a species of nematodes, *Caenorhabditis elegans* (Vanfleteren et al., 1998). The results confirm hundreds of previous studies with a wide range of species, especially those with inbred rodents housed under apparently identical but less well-controlled environments. One observes wide variations in lifespan in all these studies. For the *C. elegans* experiments the distributions of lifespan fit best with two-parameter or three-parameter logistic models and not with the classical Gompertz model or the Weibull model.

Many mutations have been shown to substantially increase lifespan in *C. elegans*. It is of interest, however, that the ranges of the lifespan variations among such mutant strains overlap with those of wild type strains (Kirkwood and Finch, 2002). Many of these long-lived mutant strains exhibit enhanced resistance to a variety of stressors, notably heat shock. It was therefore predicted that variable degrees of response to heat shock stress might form a basis, or a partial basis, for individual variations in longevity. An initial set of experiments demonstrated that is indeed the case, at least for a transgenic construct that includes the promoter of a small heat shock gene (Rea et al., 2005). There was a very strong correlation between the response to heat stress and longevity, with good-responding worms living longer. Strikingly, this phenotype was not heritable. The progeny of a worm showing a strong heat stress reaction exhibited the broad distribution of lifespan shown by the starting population. The heat stress reaction was therefore stochastic. The nature of the chance events that determine the reaction remains unknown. They could be related to the intrinsic instability of the transgene, making it important to repeat such experiments utilizing endogenous genes as reporters of the response to heat shock and other stressors. It could be due to epigenetic drifts in gene expression, perhaps involving random changes in gene promoters or in the state of chemical modifications to histone proteins that coat chromosomes. Such changes have indeed been observed in aging human identical twins (Fraga et al., 2005). While those changes have been interpreted as being driven by the environment, one cannot at present rule out random variations unrelated to environmental influences.

Variations in gene expression in genetically identical organisms examined under environmentally identical conditions have also been attributable to intrinsic noise in fundamental molecular processes such as the transcription and translation of genes. Most such observations have been made using microorganisms (Elowitz et al., 2002), but stochastic bursts of transcription have also been noted in mammalian cells (Raj et al., 2006). Moreover, substantial variations in the levels at which genes are transcribed has been shown to occur in mouse tissues, and that variation was shown to increase with age (Bahar et al., 2006).

Chance events are also of major significance in the determination of diseases of aging. For the case of cancer, mutations have been shown to be of major importance. A likely key to malignancy, however, is the chance event of suffering a mutation in a gene that when mutated, now greatly enhances the general frequency of mutation. Such genes are referred to as "mutator genes" (Bielas et al., 2006). Chance events can make the difference between life and death of individuals with coronary artery atherosclerosis, as mortality often follows the rupture of an atherosclerotic plaque, an event that is likely to be due in part to a chance event (a trigger) leading to the rupture (Falk, 1992). Moreover, some genetic interventions that have been introduced into model organisms (nematodes, mice) increase mean but not maximum lifespan and appear to rectangularize the lifespan curve. A recent example is a mouse strain carrying extra copies of a tumor suppressor locus (Matheu et al., 2007). As expected, these mice are remarkably cancer free. Of particular interest, though, their mean but not maximum lifespan was extended. Does this locus and similar interventions rectangularize the lifespan curve by reducing random events?

Initial Challenges to Consider

- What experiments might be designed in model organisms to probe the role of variations of endogenous gene expression at birth in the determinations of the remarkable stochastic variations in lifespan among genetically identical organisms?
- Which subset of endogenous genetic loci are major contributors to such stochastic variation?
- Are these variations in gene expression attributable to specific molecular events, such as chemical modifications to DNA CpG islands or histones?
- What are the molecular and biophysical mechanisms that lead to transcriptional bursts in gene expression?
- Do cells within an individual organism differ in their susceptibility to stochastic fluctuations in gene expression? For example, are postmitotic neurons more or less susceptible than cells that are destined to die or cells that turn over?

- Are there species-specific differences in the degree to which stochastic fluctuations in gene expression occur (in similar cell types)?
- Has evolution shaped the above stochastic variations in gene expression, are they adaptive, and what are the selective pressures that led to such adaptations? How can one test the hypothesis that different degrees of stochastic variations in gene expression do in fact evolve and that they are adaptive?
- To what extent do early environmental influences in developing humans (fetal, neonatal, childhood, pubertal) determine patterns of gene expression and patterns of aging in human subjects?
- Do genetic interventions that increase mean but not maximum lifespan, and appear to rectangularize the lifespan curve, act by reducing random events? Can we learn about the cellular and molecular bases for stochastic variation by testing the hypothesis that some of these interventions act by this mechanism (reducing stochastic variation)?
- What are some candidate environmental agents and social influences responsible for such putative influences and how can their impacts upon public health be measured?

Initial References

Bahar, R., C. H. Hartmann, K. A. Rodriguez, A. D. Denny, R. A. Busuttil, M. E. Dolle, R. B. Calder, G. B. Chisholm, B. H. Pollock, C. A. Klein, and J. Vijg. 2006. Increased cell-to-cell variation in gene expression in ageing mouse heart. Nature 441:1011-1014.

Bernhard, D., C. Moser, A. Backovic, and G. Wick. 2007. Cigarette smoke—an aging accelerator? Experimental Gerontology 42(3):160-165.

Bielas, J. H., K. R. Loeb, B. P. Rubin, L. D. True, and L. A. Loeb. 2006. Human cancers express a mutator phenotype. Proceedings of the National Academy of Sciences U.S.A. 103:18238-18242.

Elowitz, M. B., A. J. Levine, E. D. Siggia, and P. S. Swain. 2002. Stochastic gene expression in a single cell. Science 297:1183-1186.

Falk, E. 1992. Why do plaques rupture? Circulation 86:III-30-III-42.

Fraga, M. F., E. Ballestar, M. F. Paz, S. Ropero, F. Setien, M. L. Ballestar, D. Heine-Suner, J. C. Cigudosa, M. Urioste, J. Benitez, M. Boix-Chornet, A. Sanchez-Aguilera, C. Ling, E. Carlsson, P. Poulsen, A. Vaag, Z. Stephan, T. D. Spector, Y. Z. Wu, C. Plass, and M. Esteller. 2005. Epigenetic differences arise during the lifetime of monozygotic twins. Proceedings of the National Academy of Sciences U.S.A. 102:10604-10609.

Herskind, A. M., M. McGue, N. V. Holm, T. I. Sorensen, B. Harvald, and J. W. Vaupel. 1996. The heritability of human longevity: A population-based study of 2872 Danish twin pairs born 1870-1900. Human Genetics 97:319-323.

Kirkwood, T. B., and C. E. Finch. 2002. Ageing: The old worm turns more slowly. Nature 419:794-795.

Matheu, A., A. Maraver, P. Klatt, I. Flores, I. Garcia-Cao, C. Borras, J. M. Flores, J. Viña, M. A. Blasco, and M. Serrano. 2007. Delayed ageing through damage protection by the Arf/p53 pathway. Nature 448:375-379.

McClearn, G. E., B. Johansson, S. Berg, N. L. Pedersen, F. Ahern, S. A. Petrill, and R. Plomin. 1997. Substantial genetic influence on cognitive abilities in twins 80 or more years old. Science 276:1560-1563.

Raj, A., C. S. Peskin, D. Tranchina, D. Y. Vargas, and S. Tyagi. 2006. Stochastic mRNA synthesis in mammalian cells. PLoS Biology 4:e309.

Rea, S. L., D. Wu, J. R. Cypser, J. W. Vaupel, and T. E. Johnson. 2005. A stress-sensitive reporter predicts longevity in isogenic populations of *Caenorhabditis elegans*. Nature Genetics 37:894-898.

Reed, T., and D. M. Dick. 2003. Heritability and validity of healthy physical aging (wellness) in elderly male twins. Twin Research 6:227-234.

Vanfleteren, J. R., V. A. De, and B. P. Braeckman. 1998. Two-parameter logistic and Weibull equations provide better fits to survival data from isogenic populations of *Caenorhabditis elegans* in axenic culture than does the Gompertz model. Journals of Gerontology A—Biological and Medical Sciences 53:B393-B403.

Due to the popularity of this topic, two groups explored this subject. Please be sure to explore the other write-up, which immediately follows this one.

TASK GROUP MEMBERS—GROUP A

- Diddahally Govindaraju, Boston University
- Stephanie Lederman, American Federation for Aging Research
- Kyongbum Lee, Tufts University
- Richard Mayeux, Columbia University
- Saira Mian, Lawrence Berkeley National Laboratory
- Chris Schaffer, Cornell University
- Nicholas Schork, Scripps Genomic Medicine, Scripps Health
- Rob Stephenson, Emory University
- David Stopak, The National Academies
- Richard Suzman, National Institute on Aging
- Woodring Wright, University of Texas Southwestern Medical Center
- Alissa Poh, University of California, Santa Cruz

TASK GROUP SUMMARY—GROUP A

By Alissa Poh, Graduate Student, Science Writing Program, University of California, Santa Cruz

Boundaries between the different sciences are gradually being torn down as interdisciplinary research and open laboratories become increasingly popular concepts. Similarly, we once regarded disease and aging as independent mechanisms, but have since realized that they are in fact closely intertwined; diseases with strong age-related incidences are likely to have a strong age-associated component.

One key issue for those studying the biology of aging is that while interspecies variations in healthspan and lifespan—for example, between mice and humans—can be fully explained by genetics, the same is not true for variations *within* species. Here stochastic, or random, events are thought to be an important factor, but precious little is actually known about the role of chance in determining lifespan.

This group was hence charged with examining stochastic variation's potential influence on human lifespan. Very early in the discussion, however, it became clear that many group members were uncertain that investigations in this area should be a priority, and spent much of the first day debating its importance.

Stochastic Events and Longevity—How Much Should We Care?

"It's an enormously interesting and informative biological question," said Woodring Wright of stochastic variation itself. "But when talking about longevity, I think it's an *un*interesting question." He then offered an analogy using restaurant glasses.

"If you look at the rate of breakage of restaurant glasses, you get curves that mimic those in lifespan studies," he said. "If you have a well-built glass with a thick wall versus a flimsy wall, you'll get differences in lifespan, but there's still going to be this variation. I don't think it's an interesting question, what's responsible for the variation between when the first glass and the last one breaks, in terms of understanding the glass's lifespan."

Nevertheless, the group decided that the consequences of stochastic factors on lifespan might be sufficiently pronounced to justify developing an experimental model addressing the issue. This would be no run-of-the-mill investigation either, since the group felt that most current human longevity studies have yielded insights into factors that increase or decrease lifespan only marginally—smoking cessation, dietary changes, and drug treatments, to name a few.

"We're not looking to increase lifespan by a couple of years; rather, we need to ask what could be radically tweaked to dramatically extend lifespan," said Nicholas Schork.

A brief debate ensued around the definition of stochastic factors. The group eventually agreed that these involve subtle, random perturbations of an individual's cellular and molecular physiological milieu that accumulate over time, gradually affecting function at the whole-organism level.

The powwow then moved on to various ways to set up this experiment. Mathematical modeling was considered, where one could make statistical predictions about the impact on human lifespan of manipulating factors possibly associated with longevity. However, such modeling would only be as good as the empirical studies on which they were based and which have not proved particularly useful to date.

Taking a different approach, each group member was asked the question: "What reasonable experiments would you pursue, given a blank check?"

Genetic screens involving model organisms were suggested. The group acknowledged these to be incredibly valuable, although some members were skeptical about the potential for human genetic screens via genomewide association studies. However, such screens are unable to capture many factors implicated in the aging process.

An argument for comparative physiology and genomic studies across species was made, but because there are many species, or lineage-specific factors at play in aging and senescence, the translatability of cross-species findings would always be in doubt.

The group toyed with interventional studies, but these were also thought to be problematic. The intervention does not always work: they are only as good as the biological insights motivating the intervention. Where humans are concerned, the results are hardly radical, usually affecting lifespan in terms of only a few years.

Another suggestion was to look at contrasts between long-lived and short-lived individuals within a species. Such studies could simultaneously investigate many factors, both genetic and nongenetic, but might produce species-specific insights.

So each strategy had its drawbacks. However, by the end of the first day, the group settled on designing a study comparing individuals undergoing early and late senescence as a result of stochastic factors. This model would probably not reveal mechanisms leading, upon artificial manipulation, to dramatic lifespan differences. Still, the group felt that it would shed light on the potential role of stochastic factors in aging.

Fine Tuning the Experimental Model

Much of the second day was spent refining this proposed experiment. The group agreed that the nematode worm *C. elegans* would be an ideal test subject, since it has a very small number of cells (less than 1000), is easily manipulated, has a short lifespan, and much is already known about its biology. In particular, it has an interesting marker for senescence that can be exploited: before these worms die, they exhibit an identifiable change in swimming behavior. Researchers could thus readily access animals in the final stages of life while they are still relatively fit.

"You don't just want to pick up a dead worm and look at its gene expression profile; it's too late," Chris Schaffer commented. Or, as another group member put it, "Death is a lousy endpoint for measuring aging."

Among the group members, Wright remained openly skeptical about exploring stochastic variation's role in lifespan. He did not take issue with the feasibility of any particular experiment; rather, he questioned whether time, money, and effort should be spent in this area, or if there were worthier aspects of aging biology that should be studied first. The group therefore decided to make him the official spokesman for delivering their conclusions.

Why Genetic Screens Aren't the Last Word

The group reiterated that genetic screens in model organisms are tremendously important, as they reveal evolutionarily conserved pathways, and thus important processes, in humans. These screens do, however, miss many categories of important mechanisms.

For starters, aging is multifactorial, so the effects of individual pathways are minimal. While genetic screening has successfully identified pathways regulating multiple others downstream, single pathway effects are undetected and left by the wayside. Genetic screening also identifies processes active in postmitotic organisms that are shared by their mitotic counterparts, but additional processes could be involved in mitotic organisms. Finally,

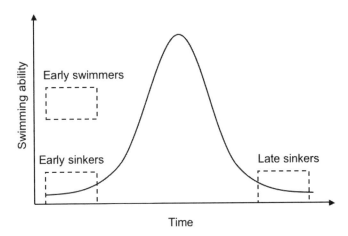

FIGURE 1 Lifespan distribution of *C.elegans* populations.

it is not possible to identify potentially significant nongenetic effects with such screens.

Among the examples of mechanisms these screens miss are the role of stem cells in aging, the importance of neoteny in the evolution of human lifespan, and most significantly, the key issue this group was asked to explore, with longevity in mind: stochastic variation in gene expression.

The Game Plan in a Nutshell

As mentioned earlier, changes in swimming behavior can be detected in *C. elegans* several days prior to death, allowing one to predict longevity and select individuals soon destined to die. This fact was exploited in the proposed model, where 10 percent of the shortest-lived worms (early sinkers) in an otherwise genetically homogeneous *C. elegans* population, as well as 10 percent of the longest-lived cohort (late sinkers), would be identified shortly before death. Healthy worms (swimmers) isolated at an early time point would serve as a control group. The lifespan distribution of these populations, marked by swimming ability, is illustrated in Figure 1.

In analyzing these different groups of worms, three hypotheses would be considered, with the first two pointing to two different roles for nongenetic factors in aging:

1. Stochastic epigenetic events generate phenotypic states associated with aging.

2. Increased variation among cells in a tissue induces an increased rate of aging due, for example, to loss of homeostatic capability.

3. Increased variation in gene expression at the single cell or organismal level does not explain the heterogeneity of lifespan in homogeneous genotypes (null hypothesis).

Many Roads to Rome

In the first hypothesis different phenotypes are produced by a change in state as a result of stochastic variation. So early and late sinkers would have identical global phenotypes due to both cohorts being ultimately channeled into senescence. This would be distinguishable from the early swimmers, as illustrated in Figure 2.

"The analogy here is that being healthy is like being in a little well at the top of a mountain, and a slight push knocks you out of the well," Schaffer explained. "Once you're out, you run all the way to the bottom. And when you've arrived there, it doesn't really matter how you were pushed; you got to the bottom of the hill, to a common state associated with senescence."

If this hypothesis is true, a potential secondary analysis would identify individual genes and pathways implicated in the fingerprint of senescence.

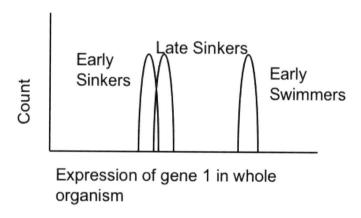

FIGURE 2 Hypothesis—"Many roads to Rome."

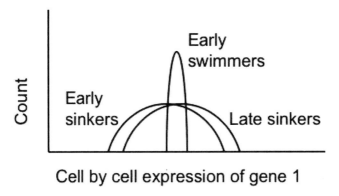

FIGURE 3 Hypothesis—"Variation drives aging."

Cross-species conservation levels of genes and pathways, as well as DNA sequence polymorphisms in human orthologs, would also be assessed.

Variation Drives Aging

All cells within a tissue start off with prescribed levels of gene expression, and any variation is small, controlled within a range where that tissue can maintain itself. In the second hypothesis, which Figure 3 illustrates, random variation in gene expression within or between specific cells drives lifespan differences. These cells become dysregulated, ultimately losing their ability to function as part of their tissue of residence. So there would be no common gene expression profile or other aspects of biology associated with senescence; rather, greater variation among cells within a tissue would result in loss of its capability.

If results indicate that variation is important, individual *genes* varying in expression from cell to cell would be identified, and vice versa (*cells* displaying cell-to-cell variation in gene expression), as secondary analyses. Cell-specific methylation patterns would be assessed, as well as cross-species conservation of the genes exhibiting greater variability. Finally, a variety of evolutionary and comparative biology experiments could be carried out on worms and flies possessing very different lifespans.

The Bigger Picture

Apart from designing this experimental model the group also pondered broader, scratch-the-surface research ideas not sufficiently refined to permit specific experiments, but nevertheless focused on developing tools to identify events critical to aging that would sail by a genetic screen. One particularly promising idea involved a series of transplant experiments: placing young tissue into an old animal and vice versa. This would help identify aging mechanisms associated with a particular cell or tissue versus organism-level processes.

In the end as the group humorously illustrated with *$$$* as a bullet point on a presentation slide, this new brand of aging-and-disease research will require a substantial financial investment to truly take wing.

"There's a real case for individual RO1s versus programmatic policies," Wright pointed out. "A lot of the things that're being found, we can't predict a priori. So there's still an important role for individual entrepreneurship to tackle isolated problems and find new handles to push, in terms of discovering new aging mechanisms."

TASK GROUP MEMBERS—GROUP B

- Suresh Arya, National Cancer Institute
- Christine Grant, North Carolina State University
- Linda Miller, *Nature*
- Richard Miller, University of Michigan
- Santa Jeremy Ono, Emory University
- Chris Patil, Lawrence Berkeley National Laboratory
- Jerry Shay, University of Texas Southwestern Medical Center
- Eric Topol, Scripps Research Institute
- Michael Torry, Steadman-Hawkins Research Foundation
- Heinz-Ulrich G. Weier, Lawrence Berkeley National Laboratory
- Iris Tse, Boston University

TASK GROUP SUMMARY—GROUP B

By Iris Tse, Graduate Writing Student, Boston University

I have cataracts, lose uphill races, get really sick when I catch the flu, girls no longer whistle when I pass by, my joints ache, and if you looked closely you'd see preclinical signs of the cancer that will kill me.

How Old Am I?

Not all species of animals age at the same rate. Humans typically will have a longer lifespan than horses, which in turn will have a longer lifespan than most rodents. However, there is a synchrony in the way age-related decline in health and diseases appears across different species. At certain points in life, for example, health problems such as vision degeneration, cancer, diabetes, central nervous system degeneration, and organ failures, will appear in some members of each species. Even more surprising is that most of these age-related diseases will appear regardless of lifespan and size of the animals. It appears as if some yet unknown species- or breed-specific factors are tying together these degenerative functional changes. Therefore, the group found this to be an area worth pursuing.

The Validity of the Stochastic Model

The multidisciplinary group initially debated in earnest the validity of the stochastic model in the context of healthspan and lifespan. The stochastic model was initially assigned to the group by the conference organizers as a springboard for discussion. Previous research has found that there is a wide range of lifespan within a genetically homogenous group. Therefore, there must be some unspecified nongenetic factor that can influence the aging trajectory and lifespan. Lifespan is not entirely controlled by genes and controllable environmental factors, and undefined gaps in current knowledge still need to be examined and studied.

However, some members of the group felt it would be premature to attribute the entirety of this unknown area to the stochastic model. The stochastic model proposed that factors extending average lifespan, but not the *maximum* lifespan, played a major role in extending healthspan. The group felt that the basis of this view was somewhat biased and would exclude many healthspan factors that also extend maximum lifespan. The stochastic model, while important, is not the complete answer.

After an hour of vigorous discussion and scrutiny, the group leader, Jerry Shay, a professor of cell biology at the University of Texas Southwestern Medical Center, articulated the group's thinking.

"To think that aging is stochastic, and therefore not within our control, greatly diminishes the impact and window of ability to understand, change, and manipulate the process of aging," said Shay.

Important Topics Surrounding Healthspan and Lifespan

Once the group set aside the stochastic model, the group members were free to explore other facets of aging. Linda Miller, the U.S. executive editor of *Nature* and the *Nature* journals, sparked the next round of discussion by pointing out neurodegeneration as an area worth exploring. Topics suggested by other members of the group included cell regenerative capacity, cellular replication and control, broad-spectrum genomic analysis of aging, and biomaterials used in current research.

However, the topic that seized the group's attention was the synchronicity of age-associated decline. Richard Miller, a professor of pathology at the University of Michigan, pointed out that many organs and cells fail at more or less the same time within a species. The sequence of illnesses are often synchronized across various species of animals, albeit at a different rate, but seemingly related to the lifespan of specific species. These co-morbid events, such as the onset of diabetes or cataracts, are not necessarily terminal illnesses and may not directly affect lifespan. However, they do affect healthspan. While the scientific basis of these assertions was unclear to some members of the group, the idea was thought to be meritorious. If we can understand why humans get cataracts at 60 years old and why mice get cataracts at 2 years old, we might be able to delay the onset of these symptoms, prolong healthspan, and understand aging a little better.

Miller insisted that it will be useful to find themes, or common families, of underlying cellular or molecular factors that time aging sequentially. Because of the lack of critical knowledge of the field by some members of the group, the initial goal was created to develop a new set of hypotheses for future experiments.

Experimental Approach

Miller already had a rough idea of the experimental approach and the group spent the remaining day and a half deliberating those points, including lengthy discussions regarding the potential benefits and pitfalls of the many different types of study designs that may best accomplish the group's target. An important first step is to generate a list of late-life dysfunctions plausibly related to the timing of aging. These age-related dysfunctions can be diseases, such as cancer, or they can be symptoms, such as cataracts or cognitive failure, that affect many animals.

Experts knowledgeable in aging or pathogenesis, or both, will need to be solicited to provide hypotheses. They also will be called upon to brainstorm a body of known factors, such as proteins, macromolecules, enzymatic activities, and genetic expressions that might influence healthspan. By understanding the synchrony of these factors, perhaps we will understand more about the synchrony of aging.

Since the group hoped to get the experiments started as quickly as possible, an important criterion is that these factors must be easy and practical to measure using established methodology and reagents. Assays that can be done on multiple species using existing methods and technology will be more useful for the project than those that depend on the development of new, species-specific reagents.

"Basically, you would want to be able use preexisting reagents or kits straight from a Sigma catalogue. Otherwise, you'd get mired over things like the trial and error of a new experimental design," said Chris Patil, a postdoctoral scholar of life sciences from Lawrence Berkeley National Laboratory.

The next step will be to select the proper animal models for interspecies comparison. The experiments will use healthy young adults with no age-related diseases because it is thought that the selective factors that mold maturation are not the same as aging. These animals will be evaluated to uncover patterns of protein expressions, or possibly some other cellular events, that determine the rate of aging of that particular species.

With some justification, four types of animals were chosen for this large-scale study:

1. *Primates*—For their similarity to humans.
2. *Bats*—In most animals, lifespan is directly correlated to body weight. However, bats have a long lifespan for their small body size, therefore allowing researchers to observe trends that are actually related to lifespan and not weight.
3. *Rodents*—Lifespan varies drastically across different species of rodents, ranging from 2 years for mice to 30 years for naked mole rats. The variety will allow experimenters to adjust for confounders.
4. *Birds*—The nonmammalian out-group to anchor the phylogenetic tree during statistical analysis.

Additional funding will be necessary to trap and collect these animals. Around 10 species for each clade of animals were thought to be adequate to start. Animals from the wild will be ideal, since they provide a more realistic snapshot of the aging process. But the group is also open to using captive animals, such as those from zoos, since they are likely to show species-specific cellular and biochemical traits that correlate with lifespan.

This broad-spectrum analysis is meant to generate useful hypotheses. It's not meant to be a hypothesis-testing exercise. The end product from these experiments will produce information necessary to create mechanistic hypotheses for further testing. It is important to provoke further questions and examinations. Interspecies contrast will be the first step. Once something comes up as a hit, then the next step will be intraspecies comparison.

"Focusing on the basic biology of aging, instead of approaching diseases one at a time, may actually speed up the process of learning how to postpone individual diseases." said Miller.

"This umbrella approach is a better way to probe the age-diseases nexus by exploiting the power of comparison."

TASK GROUP MEMBERS—GROUP C

- Craig Atwood, University of Wisconsin-Madison
- Miles Axton, Nature Genetics
- Rita Effros, University of California, Los Angeles
- Nan Jokerst, Duke University
- Jay B. Labov, National Academy of Sciences
- Valter Longo, University of California, Los Angeles
- Joao Magalhaes, Harvard University
- Ken Turteltaub, Lawrence Livermore National Laboratory
- Catherine Wolkow, National Institute on Aging, Intramural Research Program, National Institutes of Health
- Natalia Mackenzie, Boston University

TASK GROUP SUMMARY—GROUP C

By Natalia Mackenzie, Graduate Writing Student, Boston University

For many people, aging is taken for granted. Like an old machine, the human body starts to fail in its functions and eventually it stops working. Although aging seems to be an obvious and expected state of life, it has been difficult for scientists to achieve consensus on the characteristics of aging and why some organisms live longer than others. One thing though is clear:

Genes, the environment, and their interactions throughout life are key to understanding of aging.

As the nine scientists of task group C met for the first time, an engineer, an adviser in education, a science editor, and a chemist surrounded by five biologists got ready to discuss the roles that genes and the environment play in aging.

As the topic was quite broad, the first day felt like a warm-up. Intense brainstorming included a discussion about one of the main problems that scientists face while studying the biology of aging: the search for representative animal models. Because aging is a complex process that involves multiple body-part failures, it is difficult to find animal models that have all the different aspects associated with human aging. "All models simulate only one part of aging, so I am not sure we do have a model for studying aging," said one of the participants. They concluded that in addition to the most widely used animal models in research—worms, flies, and mice—animals, such as tortoises, that live much longer than humans also should be considered. This statement turned the conversation to the understanding of why some animals live longer than others. Following this idea, they discussed the concept that animals may live as long as they need to in order to optimize reproduction. In addition, they proposed that as a consequence of food scarcity, some animals may have to live longer in order to wait for better conditions to reproduce. Another possibility was that animals exposed to less predation could reproduce later and therefore live longer. But why do animals have to age at all? "To avoid competing with our kids," said one member of the group. Another participant thought that "we have to recycle matter, we don't have infinite mass."

During the second round of discussion in the first day, the group focused on the characteristics of aging and the biological and environmental events that cause it. Aging can be seen as the progressive deterioration of the organism by the accumulation of mutations and epigenetic alterations in the DNA over time. In other words, the longer we live, the more time our DNA diverges from what it was at the beginning of life. Such DNA alterations are generated by environmental factors such as sunlight (UV), ionizing radiation, pathogens (bacteria, viruses), carcinogens, replication errors, among others.

Another crucial environmental factor that affects aging is socioeconomic status (SES). SES depends on family income, parental education level, and social status in the community. It may be also understood as quality of life, considering nutrition, exposure to violence, self-esteem, and stress.

One key idea proposed as one of the main causes of animal aging was the progressive decline in the ability of cells to communicate with one another. Organs are made of different kinds of cells. For organs to properly function, the cells have to be able to converse with one another in order to synchronize growth and function. While we age, accumulation of DNA mutations and epigenetic changes generated by environmental factors affect the ability of cells to communicate with one another and therefore organs begin to fail. One good example is brain cells. The brain is composed of billions of cells that electrochemically communicate with one another. If cells are damaged and are not able to communicate, the chain of transmission is interrupted and information is lost. The result may be a decrease of memory in older people, or slower muscle reactions. According to one member of the group, epigenetic DNA changes, such as histone methylation and DNA CpG methylation, accumulate and respond faster to environmental changes than mutational damage to DNA.

Fortunately, there are intrinsic cellular mechanisms that patrol the integrity of the DNA. If the repairing machinery detects a change, it can repair the mistake from within the cell. Therefore, how long an animal lives also depends on the efficiency of its repairing machinery. However, at some point in life, not even the repairing machinery can save the body from deteriorating, because mutation accumulation can also affect the integrity of the repairing machinery itself. As a consequence, mistakes are no longer repaired and an overwhelming accumulation of altered DNA damages or kills the cells. Another flaw of the repairing machinery is that it cannot compare genome sequence between individual cells in order to "homogenize" their genetic information after they start accumulating differences. It was proposed that the way life is designed to reset all this damaged information is by eliminating the old organism, generating a whole new one by fertilization of the egg by the sperm.

During the second day, the group focused on building a tangible proposal that would represent their ideas and conclusions. They first decided that their working hypothesis would be that aging results from increasing cellular damage that compromises communication pathways, leading to impaired cellular functioning and organ failure. The group felt that there was currently a great deal of research and information about aging at the cellular and organismal levels, but that there remain large gaps of knowledge in between (e.g., at the tissue, organ, and organ system levels). According to the group, one reason for that was that scientists studying different aspects of aging do not communicate enough with one another. Agreeing with one member's comment that "in the intersection among different points of view is where creativity lies," the group decided to propose an interdisciplinary research initiative on aging that would begin to break down miscommunication among disciplines and mine data from research on aging at the cellular and organismal levels to offer possible new insights into the process at additional levels of biological organization.

Their final proposal was a five-year program that would focus on creating biologically guided tools that help extend life- and healthspan. The idea emerged when the only engineer in the room said that engineers only need to know "what is broken" in aging so they can fix it. "We don't know what you biologists want," said the engineer. The group immediately understood how powerful the fusion of both disciplines could be for contributing not only to aging research but also to science in general.

The main goal of the proposed program would be to use biology to guide machines that would allow early detection and prevention of genetic and epigenetic modifications that influence aging. The idea is to use these new tools to intervene in the aging process by repairing, reprogramming, removing, or replacing damaged biological components of cells that contribute to the deterioration of the human body. Specifically, they proposed to focus on organ systems that are mostly affected in aging like vasculature, the nervous system, musculoskeletal, immune, vision, and hearing.

The group decided that the first year should be oriented to data mining, the integration of all the existing knowledge of aging at a cellular, tissue, organ, and organism level. The next two to five years, experiments and technology development would take place where the most promising research would be prioritized and engineering prototype devices would be made.

One example of a tool that the group proposes relates to hormonal imbalances that normally inhibit physical, sexual, and cognitive functions as a result of aging. The idea is to first identify those hormones that are altered and create devices to monitor and redirect hormones and behavior to what is observed in younger organisms.

But the conclusion this task group presented the last day of the conference was not only a very practical one. They were also able to call the scientific community's attention to breaking down the silos and to moving to a multidisciplinary approach in aging research.

Inflammation's Effects on Aging

TASK GROUP DESCRIPTION

Background

Inflammatory processes are recognized in many chronic conditions that alter outcomes of health during aging, including atherosclerosis; dementias of the Alzheimer type; cancer; diabetes; chronic obstructive pulmonary disease; obesity; autoimmune disorders; and chronic viral, bacterial, and fungal diseases. Environmental factors include infectious agents and inflammogens from air and diet. Conversely, clinical, experimental, and epidemiological lines of evidence have shown that anti-inflammatory drugs attenuate some of these conditions. In part, these synergies have been recognized in the National Institutes of Health director's Gene, Environment, and Health Initiative (GEI) and in the National Institute of Environmental Health Sciences and National Human Genome Research Institute programs in environmental genomics. It seems timely and appropriate in the context of this conference to have a session on inflammation and aging. The scope should include human genomics and population diversity, multigenerational effects, and changing ecological factors. As urban populations continue to grow and as water and air quality deteriorate globally, we may anticipate increasing global exposure to infection and inflammation. The consequences to health during aging of the growing inflammatory burden have not been well articulated and new experimental approaches may be needed.

Initial Challenges to Consider

- What are the impacts of specific pro-inflammatory agents during the fetal, neonatal, childhood, pubertal, and reproductive periods of the life course on health during the last half of the life course?
- How will more discussion be developed between researchers and policy makers on biomedical interventions to aging and the industrial-ecological issues of air and water pollution?
- Will novel new agents evolve that are comparable to the late-life deleterious effects of microbial agents, such as type A beta-hemolytic streptococci or cigarette smoke?
- What are the trade-offs with the long-term administration of many pharmacological interventions, including anti-inflammatory agents? For example, low-dose aspirin can lead to fatal gastrointestinal or cerebral hemorrhages in susceptible individuals.
- Will physicians be able to take advantage of new genomic methodologies to predict who will or will not be at high risk for such side effects? More generically, can algorithms be developed to arrive at rational conclusions with regard to risk assessments for individuals and cost-benefit analyses for the case of populationwide interventions?
- Pharmaceutical companies are at high risk for costly lawsuits involving unanticipated serious complications of new drugs and vaccines, the development of which requires investments of millions of dollars. These are among the factors that discourage the development of new agents for infectious agents that are rare in the United States but are common in undeveloped countries. What business models (e.g., partnerships between industry and government) would address this problem?

Initial References

Antonicelli, F., G. Bellon, L. Debelle, and W. Hornebeck. 2007. Elastin-elastases and inflamm-aging. Current Topics in Developmental Biology 79:99-155.

Cauley, J. A., M. E. Danielson, R. M. Boudreau, K. Y. Forrest, J. M. Zmuda, M. Pahor, F. A. Tylavsky, S. R. Cummings, T. B. Harris, and A. B. Newman. 2007. Inflammatory markers and incident fracture risk in older men and women: The health aging and body composition study. Journal of Bone and Mineral Research 22(7):1088-1095.

Crimmins, E. M., and C. E. Finch. 2006. Infection, inflammation, height, and longevity. Proceedings of the National Academy of Sciences U.S.A. 103:498-503.

Finch, C. E. 2005. Developmental origins of aging in brain and blood vessels: An overview. Neurobiology of Aging 26:281-291.

Finch, C. E. 2007. *The Biology of Human Longevity: Inflammation, Nutrition, and Aging in the Evolution of Lifespans*. Burlington, Mass.: Academic Press/Elsevier.

Finch, C. E., and T. E. Morgan. 2007. Systemic inflammation, infection, ApoE alleles, and Alzheimer disease: A position paper. Current Alzheimer Research 4(2):185-189.

Gavilán, M. P., E. Revilla, C. Pintado, A. Castaño, M. L. Vizuete, I. Moreno-González, D. Baglietto-Vargas, R. Sánchez-Varo, J. Vitorica, A. Gutiérrez, and D. Ruano. 2007. Molecular and cellular characterization of the age-related neuroinflammatory processes occurring in normal rat hippocampus: Potential relation with the loss of somatostatin GABAergic neurons. Journal of Neurochemistry 103(3):984-996.

Griffiths, M., J. W. Neal, and P. Gasque. 2007. Innate immunity and protective neuroinflammation: New emphasis on the role of neuroimmune regulatory proteins. International Review of Neurobiology 82:29-55.

Vasto, S., G. Candore, C. R. Balistreri, M. Caruso, G. Colonna-Romano, M. P. Grimaldi, F. Listi, D. Nuzzo, D. Lio, and C. Caruso. 2007. Inflammatory networks in ageing, age-related diseases and longevity. Mechanisms of Ageing and Development 128(1):83-91.

Task Group Members

- Abraham Aviv, University of Medicine and Dentistry of New Jersey
- Shea Gardner, Lawrence Livermore National Laboratory
- George A. Kuchel, University of Connecticut
- Sarah Kummerfeld, Stanford University
- J. Christopher Love, Massachusetts Institute of Technology
- Helen Vlassara, Mount Sinai School of Medicine
- Mary White, Centers for Disease Control and Prevention
- Allyson Collins, Massachusetts Institute of Technology

TASK GROUP SUMMARY

By Allyson Collins, Graduate Student, Science Writing, Massachusetts Institute of Technology

"Aging begins at fertilization," said Caleb Finch, codirector of the University of Southern California's Alzheimer's Disease Research Center and professor in the neurobiology of aging. The statement illustrated the enormity of the task for the seven others around the table—scientists, geriatricians, a public health official, and a chemical engineer—charged with evaluating inflammation's effect on aging. Finch, who inspired the topic selection, initiated conversation about the subject on day one of the

National Academies Keck *Futures Initiative* Conference, and then left the group members to tackle the topic.

Narrowing the Problem

The first issue: oxidative stress (OS), the body's inability to control high levels of cell-damaging reactive oxygen that is produced by metabolism. The process of aging in most mammals includes an increased burden of oxidative stress and inflammation, as well as a declining innate immunity. In humans, high OS and inflammation are both involved in many diseases of aging, including atherosclerosis, arthritis, cancer, Parkinson's, and Alzheimer's. But several group members asked what causes this burden on the body? Is it related to the environment, is it the result of an intrinsic problem, or is it just the nature of aging?

These fundamental questions led George Kuchel, chief of the Division of Geriatric Medicine and director of the UConn Center on Aging at the University of Connecticut, to ask: "Do we know that elevated peripheral inflammatory markers, which have been shown to predict disability, reflect the presence of tissue inflammation? Does this process drive the progression of disease, frailty, and disability? Should we be trying to eliminate inflammation, or does it actually reflect the body's normal compensatory mechanism to injury?" These uncertainties remained an underlying thread in much of the discussion, and were soon joined by a host of others: From where do high oxidative stress and inflammation originate? Does high OS/inflammation cause age-related disease, or does it result from disease? At what age do the OS/inflammatory responses emerge?

It's difficult to pin down both the source of oxidative stress and inflammation as well as their effects on the body because multiple issues are involved. When older adults present to their physicians, their problems cross the boundaries of organs, and inflammation is just one such systemic problem. Because this stress likely accumulates over many years, no explicit formula exists for calculating the increasing trend of oxidative stress and inflammation with age. Many interacting aspects contribute, including the endogenous factors such as metabolic and hormonal changes, gender, race, and genetic variation, and the exogenous factors such as diet, physical activity, environmental pollutants or irritants, socioeconomic status, and stress. Therefore, by reducing or eliminating these causes, age-related diseases could also decline, possibly resulting in an increased healthspan and ultimately a longer lifespan.

The group agreed on the complexity of the causes of OS/inflammation, and also on the dramatic changes in external stimuli over the past 30 to 50 years. Modern diets are higher in advanced glycation end products (AGEs), harmful compounds produced after consuming heated, sterilized, or processed foods, which may significantly contribute to oxidative stress and chronic diseases. Also, changes in our culture have brought about transformations in patterns of physical activity, socioeconomic status, stress, chronic infections, and the environment, all of which increasingly affect inflammation.

As the day progressed, ideas continued to flow, but group members struggled in pinpointing and defining the challenges involved. They also expressed frustration with the narrowness of the topic. "We can redefine the questions as a group," said Helen Vlassara, the appointed leader, and professor and director of the Division of Experimental Diabetes and Aging at Mount Sinai School of Medicine. "We've been given license for that." Yet no one dared to propose an alternative. So the conversation continued but directed more toward possible remedies.

Identifying the Solutions

First, Abraham Aviv, professor and director of the Center of Human Development and Aging at the University of Medicine and Dentistry of New Jersey, approached the subject of low-grade inflammation in relationship to aging conditions such as cancer and cardiovascular disease. "Wouldn't it be a great idea if we could find anti-inflammatory drugs that don't have the side effects and could impact these diseases?" he asked. Then, Sarah Kummerfeld, a postdoctoral fellow at Stanford University spoke up. "We have an inbuilt system to upregulate antioxidants. Can we make that start working harder to induce the antioxidant system?" she suggested, in an effort to combat the oxidant stress.

And later, Kuchel attempted to summarize the conversation about inflammation therapies: "They can be highly targeted interventions, targeting specific organs, or they can be very broad, such as dietary manipulations, exercise, or improved living standards. Those are attractive precisely because they are pleotropic," or have multiple effects, he said. Then, Mary White, an epidemiologist and branch chief in the Division of Cancer Prevention and Control at the Centers for Disease Control and Prevention, asked whether the scientific knowledge was sufficient to support population-based interventions, such as fortifying processed foods with a compound that could combat inflammation, or altering messages about a healthy diet to address oxidative stress.

But in the midst of the ideas, Vlassara mentioned that the base level of inflammation in humans might actually need to be redefined due to overexposure to oxidants. "The earlier we diagnose that, the earlier we can have an impact on decision making," she said. So they changed gears and began considering long-term, population-based studies to track changes in factors related to inflammation over time. However, these studies require a well-defined insult to the system and a well-defined bodily reaction to track the physiological changes associated with oxidative stress. In addition, such a study done in a prospective manner might require 50 years or more to gain results.

By the end of the day the group had conferred about many ideas and possible solutions relating to the topic that they had previously considered to be narrow. When they regrouped the following morning, Vlassara announced the results of her PubMed search the previous night—more than 21,000 citations on inflammation and aging, indicating the overwhelming evidence on the subject. "We need to start coming to an agreement," Kuchel said, and the topic turned to the role of gender in inflammation. Why is it that estrogen in most cases produces anti-inflammatory effects, but in some situations exhibits pro-inflammatory behavior? Approaching this issue in terms of gender naturally leads to questioning the effects of hormones on innate immune responses. The group proposed the naked mole rat, which doesn't undergo menopause, as an animal model for studies of hormone effects and gender differences in aging inflammatory responses.

Aging animal models could also be the means for clarifying the cause-and-effect relationship between inflammation and age-related diseases. Through these models, disease burden could be linked with levels of inflammatory factors, internal and external oxidant pressures could be studied, and profiles of OS/inflammatory changes could be established. Information gained from these types of studies could be used to predict the onset of disease. "Our priority needs to be linking these factors to disability," Kuchel noted. "We're not here to address a specific disease or condition. We're here to address what we can do to increase people's health and functional independence as they age."

So the group returned to human studies and the topic of diet. They suggested that additional research could focus on cross-sectional, longitudinal analyses of AGE-restricted diets, and diet's effect on chronic inflammatory diseases. The results could also affect nutritional policies in the future—guidelines could be instituted for disclosing AGE content in food products, and limits could be imposed on the total amount of AGEs allowed.

A diagnostic tool that might be used to test for reactive oxygen species and AGEs is called an enzyme-linked immunosorbent assay, or ELISA. High-throughput screening methods could also be put into place to detect inflammatory markers and DNA mutations induced by particular inflammogens. Further analysis could involve the relationship of these factors to problems with blood vessels, loss of muscle mass, and general age-related weakness.

Another hot topic became the telomeres at the tips of chromosomes. Aviv explained to the group that in white blood cells, the length of telomeres decreases as the cumulative burden of oxidative stress in the organism increases. Future research could clarify the relationship between inflammogens and the length and functioning of telomeres in the aging immune system.

Next, Aviv suggested that the group consider developing diagnostic tests to distinguish between inflammation and oxidative stress, and to analyze biomarkers linked to the conditions. This gave Kuchel an idea: establishing an OS/inflammation equivalent of Koch's postulates, which in 1890 outlined the criteria for determining that a specific infectious organism caused a disease. This, however, has been determined in many studies in animals and several in humans.

Discussing the Limitations

With the conference winding down, the group began considering the limits of current technology that would need to be overcome before implementing these solutions. First, the members noted that technology for assessing inflammatory markers in the clinic is unavailable. Physicians need low-cost, minimally invasive, rapid methods for analyzing reactive oxygen species and AGEs that could be performed during annual physicals. This type of system could even lead to home-based monitoring of OS markers, similar to diabetic kits for tracking glucose levels. In addition, a comprehensive database of genetic, protein, and metabolic data from a diverse human population is also lacking.

Near the end, Finch, the inspiration behind the topic, returned. "Our effort . . . can stimulate general discussion on the topic," he said, offering encouragement to the smallest group at the conference, its participants mixed with both basic and clinical research backgrounds. Each member offered a fresh perspective on the topic of inflammation and aging, and each left with a detailed list of questions and a variety of solutions that may not have been conceived without this interdisciplinary experience.

Appendixes

Preconference Webcast Tutorials

**September 21, 2007, 1:00-4:00 p.m. EDT
(10:00 a.m.-1:00 p.m. PDT)**

Demography of Aging: Recent and Expected Trends in Human Life Expectancy

Ken Wachter
Professor of Demography and Statistics
Chair, Department of Demography
University of California, Berkeley

Stress, Lifestyle, and Prevention of Decline

Teresa Seeman
Professor
Division of Geriatrics, Department of Medicine
University of California, Los Angeles

Quality of Life Technology

Rory Cooper
Distinguished Professor
FISA/PVA Chair of Rehabilitation Science and Technology
University of Pittsburgh

**September 25, 2007, 1:00-3:30 p.m. EDT
(10:00 a.m.-12:30 p.m. PDT)**

Demography of Aging: Recent and Expected Trends in Functional Status and Active Life Expectancy in Late Life

Vicki Freedman
Professor
School of Public Health
University of Medicine and Dentistry of New Jersey

Gerontology 101/Cellular and Molecular Causes of Aging

Caleb Finch
ARCO-William F. Kieschnick Professor in the Neurobiology of Aging
Co-Director, Alzheimer Disease Research Center
University of Southern California

**September 26, 2007, 1:00-4:00 p.m. EDT
(10:00 a.m.-1:00 p.m. PDT)**

Regenerative Medicine: Prolonging Life through Replacement, Repair, and Regeneration

Robert Nerem
Parker H. Petit Distinguished Chair for Engineering in Medicine
Institute Professor and Director of the Parker H. Petit Institute for Bioengineering and Bioscience
Georgia Institute of Technology

Animal Models in Aging Research

Steve Austad
Professor of Cellular and Structural Biology
The University of Texas Health Science Center at San Antonio

Social and Behavioral Determinants of Healthy Life Expectancy

Eileen Crimmins
Edna M. Jones Professor of Gerontology
University of Southern California

Agenda

November 14-16, 2007

Wednesday, November 14, 2007

7:15 and 7:45 a.m.	Bus pickup (From the Hyatt Regency Newport Beach to the Beckman Center
7:30 a.m.	Registration (Beckman Center/Outside Auditorium)
7:30-8:30 a.m.	Breakfast (Beckman Center/Dining Room)
8:30-9:00 a.m.	**Welcome and Opening Remarks** (NAS, IOM, and NAE Presidents, Steering Committee Chair, Keck Foundation) (Beckman Center/Auditorium)
9:00-9:30 a.m.	**Keynote Address** (Auditorium) Michael M. Merzenich Francis Sooy Professor of Otolaryngology Keck Center for Integrative Neurosciences University of California, San Francisco, School of Medicine

9:30-10:30 a.m.	**Panel Discussion** (Auditorium) (Q&A with Webcast Tutorial Speakers) *Moderator* Jack Rowe, Professor, Mailman School of Public Health, Columbia University *Panelists* • Steven Austad, Professor of Cellular and Structural Biology, The University of Texas Health Science Center at San Antonio • Rory Cooper, Distinguished Professor, FISA/PVA Chair of Rehabilitation Science and Technology, University of Pittsburgh • Eileen Crimmins, Edna M. Jones Professor of Gerontology, University of Southern California • Caleb Finch, ARCO-William F. Kieschnick Professor in the Neurobiology of Aging; Co-Director, USC Alzheimer Disease Research Center, University of Southern California
10:30-11:00 a.m.	Break (Atrium)
11:00 a.m.-noon	**Panel Discussion Continued** (Auditorium) (Q&A with Webcast Tutorial Speakers) *Moderator* Jack Rowe, Professor, Mailman School of Public Health, Columbia University *Panelists* • Vicki Freedman, Professor, School of Public Health, University of Medicine and Dentistry of New Jersey • Robert Nerem, Parker H. Petit Distinguished Chair for Engineering in Medicine Institute; Professor and Director of the Parker H. Petit Institute for Bioengineering and Bioscience, Georgia Institute of Technology • Teresa Seeman, Professor, Division of Geriatrics, Department of Medicine, University of California, Los Angeles

	• Ken Wachter, Professor of Demography and Statistics; Chair, Department of Demography, University of California, Berkeley
Noon-12:30 p.m.	**Task Group and Grant Program Overview** (Auditorium) (Jack Rowe)
12:30-2:00 p.m.	Lunch (Dining Room) Setup for Poster Sessions 1 and 2 (Hallways A and B)
1:15-2:00 p.m.	Poster Session 1 (Hallway A—see "Posters" tab in binder)
2:00-5:30 p.m.	Task Group Session 1 Various Meeting Rooms (see "Task Groups" tab in binder)
3:30-4:00 p.m.	Break (Atrium, Second Floor Hallway)
5:30-7:00 p.m.	Reception (Beckman Center/Fountain Courtyard)
5:45-6:30 p.m.	Poster Session 2 (Hallway B)
7:00-9:00 p.m.	Communication Awards Presentation and Dinner (Atrium)
9:00 p.m.	Bus pickup (From Beckman Center to Hyatt Regency Newport Beach)
9:00-11:00 p.m.	Informal Discussions/Hospitality Room (optional) (Hyatt Regency Newport Beach/Patio Room)

Thursday, November 15, 2007

7:00 and 7:30 a.m.	Bus pickup (From the Hyatt Regency Newport Beach to the Beckman Center)
7:15-8:00 a.m.	Breakfast (Beckman Center/Dining Room)

Time	Activity
8:00-10:00 a.m.	Task Group Session 2 (Various Meeting Rooms—see "Task Groups" tab in binder)
10:00-10:30 a.m.	Break (Atrium, Second Floor Hallway)
10:30-noon	Task Group Reports (5-6 minutes per group) (Auditorium)
Noon-1:30 p.m.	Lunch (Dining Room)
12:45-1:30 p.m.	Related Task Group Discussions (Groups 2A-2B; 6A-6B; 7A-7C) Balboa Room (First Floor), Groups 2A-2B Newport Room (First Floor), Groups 6A-6B Board Room (First Floor), Groups 7A-7C
1:30-5:00 p.m.	Task Group Session 3 (Same Meeting Room as Sessions 1 and 2)
3:00-3:30 p.m.	Break (Atrium and Second Floor Hallway)
5:00 p.m.	Task Group representatives to drop off presentation at information/registration desk (Atrium)
5:00-6:30 p.m.	Reception (Beckman Center/Fountain Courtyard)
6:30-8:00 p.m.	Dinner (Beckman Center/Atrium)
8:00-8:30 p.m.	**Dinner Speaker** (Atrium) John Rowe, M.D. Professor Mailman School of Public Health Department of Health Policy and Management Columbia University
8:30 p.m.	Bus pickup (From Beckman Center to Hyatt Regency Newport Beach)
9:00-11:00 p.m.	Informal Discussions/Hospitality Room (optional) (Hyatt Regency Newport Beach/Patio Room)

AGENDA

Friday, November 16, 2007

7:00 and 7:30 a.m.	Bus pickup (From the Hyatt Regency Newport Beach to the Beckman Center)
7:15-8:00 a.m.	Breakfast (Beckman Center/Dining Room)
7:15 a.m.	Stop by registration/information desk to arrange for taxi service if shuttle bus service at noon and 1:30 p.m. does not work with schedule (Beckman Center/Atrium/Registration and Information Desk)
8:00-9:30 a.m.	Task Group Reports (10-12 minutes per group) (Auditorium)
9:30-10:00 a.m.	Break (Atrium)
10:00-11:00 a.m.	Task Group Reports Continued (Auditorium)
11:00-noon	Q&A Across All Task Groups (Auditorium)
Noon-1:30 p.m.	Lunch (optional) (Dining Room)
Noon & 1:30 p.m.	Buses depart for Hotel and Airport (Buses depart Beckman Center for Hyatt Regency Newport Beach and John Wayne [SNA] Airport)

Participants

Steven Abramowitch
Research Assistant Professor
Bioengineering
University of Pittsburgh

Jad Abumrad
Host/Producer
WNYC

Richard Allman
Professor and Director
Center for Aging
University of Alabama at
    Birmingham

Andrea Anderson
Graduate Science Writing Student
New York University

Suresh K. Arya
Program Director and Senior
    Investigator
National Cancer Institute
National Institutes of Health

Megan Atkinson
Senior Program Specialist
Keck *Futures Initiative*
The National Academies

Craig Atwood
Associate Professor, Medicine
University of Wisconsin-Madison/
    Veterans Affairs Hospital

Steven N. Austad
Professor
Cellular and Structural Biology
The University of Texas Health
    Sciences Center at San
    Antonio

Abraham Aviv
Professor and Director
The Center of Human
    Development and Aging
New Jersey Medical School
University of Medicine and
    Dentistry of New Jersey

Myles Axton
Editor
Nature Genetics
Nature Publishing Group

Albert Banes
Professor
Biomedical Engineering
North Carolina State/University of North Carolina, Chapel Hill, Joint Department

Noah Barron
Graduate Science Writing Student
University of Southern California

Lazelle Benefield
Professor and Parry Chair in Gerontological Nursing
College of Nursing
University of Oklahoma Health Sciences Center

Allyson Bennett
Assistant Professor
Physiology and Pharmacology; Pediatrics
Wake Forest University School of Medicine

Floyd E. Bloom
Professor Emeritus
Molecular and Integrative Neuroscience
The Scripps Research Institute

Kath Bogie
Senior Reseach Associate/Senior Research Scientist
Orthopaedics
Case Western Reserve University

Bambi Brewer
Assistant Professor
Rehabilitation Science and Technology
University of Pittsburgh

Liming Cai
Senior Service Fellow
National Center for Health Statistics

Judith Campisi
Senior Scientist, Professor
Life Sciences Division
Lawrence Berkeley National Laboratory
Buck Institute for Age Research

John Cannon
Graduate Science Writing Student
Science Communication Program
University of California, Santa Cruz

James Carey
Professor
Entomology
University of California, Davis

PARTICIPANTS

Laura L. Carstensen
Professor and Vice Chair and
 Director
Psychology
Stanford University

Nadeen Chahine
Lawrence Fellow
Engineering Technologies Division
Lawrence Livermore National
 Laboratory

Megan Chao
Graduate Science Writing Student
University of Southern California

Daofen Chen
Director
Sensorimotor Integration Program
National Institute of Neurological
 Disorders and Stroke

Ralph Cicerone
President
National Academy of Sciences

Allyson Collins
Graduate Science Writing Student
Massachusetts Institute of
 Technology

Rory Cooper
Distinguished Professor, FISA/PVA
 Chair
Rehabilitation Science and
 Technology
University of Pittsburgh

Eileen Crimmins
Professor
Center on Biodemography and
 Population Health
University of Southern California

Clifford Dacso
Executive Director
Abramson Center for the Future of
 Health
The Methodist Hospital Research
 Institute

Cristina Davis
Assistant Professor
Mechanical and Aeronautical
 Engineering
University of California, Davis

Nandini Deshpande
Assistant Professor
Physical Therapy and
 Rehabilitation Sciences
University of Kansas Medical
 Center

John Doyle
John G. Braun Professor of
 Control and Dynamical
 Systems, Electrical Engineering
 and BioEngineering
California Institute of Technology

Rita Effros
Professor
Pathology and Laboratory
 Medicine
David Geffen School of Medicine
 at University of California, Los
 Angeles

Leon Esterowitz
Program Director
Division of Chemical,
    Bioengineering,
    Environmental, and
    Transport Systems
National Science Foundation

Alicia Figueiredo
Director
The Presidents' Circle
The National Academies

Caleb Finch
ARCO-William F. Kieschnick
    Professor in the
    Neurobiology of Aging and
    Co-Director
Alzheimer Disease Research
    Center
University of Southern
    California

Harvey Fineberg
President
Institute of Medicine

Dorothy Fleisher
Program Director
W. M. Keck Foundation

Richard N. Foster
Managing Partner, Millbrook
    Management Group LLC
Board Member, W. M. Keck
    Foundation

Vicki Freedman
Professor
Health Systems and Policy
University of Medicine and
    Dentistry of New Jersey

Ken Fulton
Executive Director
National Academy of Sciences

Shea Gardner
Computations
Lawrence Livermore National
    Laboratory

Diddahally Govindaraju
Associate Professor
Neurology
Boston University School of
    Medicine

Christine Grant
Professor of Chemical Engineering
Chemical and Biomolecular
    Engineering
North Carolina State University

Geoffrey Graybeal
Graduate Science Writing Student
Grady College of Journalism and
    Mass Communication
University of Georgia

X. Edward Guo
Associate Professor of Biomedical
    Engineering
Biomedical Engineering
Columbia University

PARTICIPANTS

Mary Haan
Professor
Epidemiology
University of Michigan

Jong-in Hahm
Assistant Professor
Chemical Engineering
The Pennsylvania State University

Arun Hampapur
Manager and Research Staff
    Member
Exploratory Computer Vision
    Group
IBM T. J. Watson Research Center

Stuart Harshbarger
System Integrator
National Security Technology
    Department
Johns Hopkins University Applied
    Physics Laboratory

Anne Heberger
Evaluation Research Associate
Keck *Futures Initative*
The National Academies

James Herndon
Research Professor of Neuroscience
Yerkes National Primate Research
    Center
Emory University

Richard J. Hodes
Director
National Institute on Aging
National Institutes of Health

Scott Hofer
Professor
Human Development and Family
    Sciences
Oregon State University

Brian Hofland
Program Director, International
    Aging Team
The Atlantic Philanthropies (USA)
    Inc.

Stephen Intille
Technology Director, House_n
Architecture
Massachusetts Institute of
    Technology

Robert Jaeger
Program Director
National Science Foundation

Nan Jokerst
J. A. Jones Professor of Electrical
    and Computer Engineering
Duke University

Matt Kaeberlein
Assistant Professor
Pathology
University of Washington

James Kahan
Adjunct Behavioral Scientist
RAND Corporation

Eric Kandel
Author
Columbia University

Jeffrey Kaye
Professor
Neurology and Biomedical
   Engineering
Oregon Health and Science
   University

Don Kennedy
Editor in Chief
Science

Corey Keyes
Associate Professor
Sociology, with a joint
   appointment in Public Health
Emory University

Lauren Gerard Koch
Assistant Professor
Physical Medicine and
   Rehabilitation
University of Michigan

Steven Kou
Associate Professor
Industrial Engineering and
   Operations Research
Columbia University

Bruce Kristal
Associate Professor
Department of Neurosurgery
Brigham and Women's Hospital

Robert Krulwich
Co-Host
WNYC

George A Kuchel
Travelers Chair (Geriatrics),
   Professor (Medicine)
Director, UConn Center on Aging
University of Connecticut Health
   Center

Vikram Kumar
Chief Medical Officer
Cogito Health Inc.
Brigham and Women's Hospital

Sarah Kummerfeld
Developmental Biology
Stanford University

Jay B. Labov
Senior Adviser for Education and
   Communication
Center for Education
National Academy of Sciences

Kenneth Langa
Associate Professor
General Medicine and Institute for
   Social Research
University of Michigan

Erin Lavik
Assistant Professor
Biomedical Engineering
Yale University

Stephanie Lederman
Executive Director
American Federation for Aging
   Research

PARTICIPANTS

Ronald Lee
Professor
Demography
University of California, Berkeley

Kyongbum Lee
Assistant Professor
Chemical and Biological
 Engineering
Tufts University

Sean Leng
Assistant Professor of Medicine
Division of Geriatric Medicine and
 Gerontology
Johns Hopkins University School
 of Medicine

Rachel Lesinski
Senior Program Specialist
Keck *Futures Intiative*
The National Academies

Howard Leventhal
Board of Governors Professor of
 Health Psychology
Institute for Health and
 Department of Psychology
Rutgers University

Jane Liaw
Graduate Science Writing Student
University of California, Santa
 Cruz

Su-Ju Lin
Assistant Professor
Microbiology
University of California, Davis

Valter Longo
Associate Professor, Hanson Chair
 of Biogerontology
Gerontology and Molecular and
 Computational Biology
University of Southern California,
 Los Angeles

Christopher Love
Assistant Professor
Chemical Engineering
Massachusetts Institute of
 Technology

Natalia Mackenzie
Graduate Science Writing Student
Boston University

Richard Macko
Director, Maryland Exercise and
 Robotics Center
Neurology, Medicine, Physical
 Therapy
University of Maryland School of
 Medicine/Baltimore Veterans
 Affairs Hospital

Joao Magalhaes
Genetics
Harvard Medical School

Kenneth Manton
Research Professor, Arts and
 Sciences
Duke University

George M. Martin
Professor of Pathology Emeritus;
  Director Emeritus
Alzheimer's Disease Research
  Center
University of Washington

Maja Matarić
Professor, Senior Associate Dean
  for Research
Computer Science and
  Neuroscience
University of Southern California

Richard Mayeux
Gertrude H. Sergievsky Professor
Neurology, Psychiatry, and
  Epidemiology
Columbia Univeristy

Stephen McAleavey
Assistant Professor
Biomedical Engineering
University of Rochester

Graham J. McDougall Jr.
Professor
Nursing
The University of Texas at Austin

Michael Merzenich
Francis Sooy Professor of
  Otolaryngology
Neurosciences
University of California, San
  Francisco

Saira Mian
Computer Staff Scientist
Life Sciences Division
Lawrence Berkeley National
  Laboratory

Richard Miller
Professor
Pathology and Geriatrics
University of Michigan

Linda Miller
U.S. Executive Editor
Nature and The Nature Journals
Nature Publishing Group

Duncan Moore
Professor of Optical Engineering
The Institute of Optics
University of Rochester

Russell E. Morgan Jr.
President
SPRY Foundation

Laura Mosqueda
Director, Professor
Geriatrics
University of California, Irvine,
  Medical Center

Laurence Mueller
Professor, Ecology and
  Evolutionary Biology
University of California, Irvine

Karim Nader
William Dawson Chair
Psychology
McGill University

Hamid Najib
Personal Computer and Program
 Support Specialist
The National Academies

Robert M. Nerem
Parker H. Petit Distinguished
 Chair for Engineering in
 Medicine, Institute Professor
 and Director of the Parker
 H. Petit Institute for
 Bioengineering and Bioscience
Georgia Institute of Technology

Greg O'Neill
Director
National Academy on an Aging
 Society

Santa Jeremy Ono
Vice Provost, Deputy Provost
Academic Initiatives
Emory University

Elaine Oran
Laboratory for Computational
 Physics
U.S. Naval Research Laboratory

Steven Orzack
President and Senior Research
 Scientist
Fresh Pond Research Institute

Chris Patil
Postdoctoral Scholar
Life Sciences
Lawrence Berkeley National
 Laboratory

Sara Peckham
Wellness Consultant

Hunter Peckham
Professor
Biomedical Engineering
Case Western Reserve University

Margaret Perkinson
Associate Professor
Occupational Science and
 Occupational Therapy
Saint Louis University

Marty Perreault
Director
Sustainability Roundtable
Science and Technology for
 Sustainability Program
The National Academies

Daniel Perry
Executive Director
Alliance for Aging Research

Alissa Poh
Graduate Science Writing Student
Science Communication Program
University of California, Santa
 Cruz

Alan Porter
Evaluation Coordinating
  Consultant
Technology Policy and Assessment
  Center
Georgia Institute of Technology

Ismael Rafols
Marie-Curie Intra-European
  Research Fellow
University of Sussex

Shane Rea
The University of Texas Health
  Science Center at San Antonio

James Rimmer
Professor/Director
National Center on Physical
  Activity and Disability
University of Illinois at Chicago

Dave Roessner
Evaluation Consultant
Keck *Futures Initiative*
The National Academies

Michael Rose
Professor
Ecology and Evolutionary Biology
University of California, Irvine

Corinna Ross
Postdoctoral Fellow
Barshop Institute for Longevity
  and Aging Studies
University of Texas Health Science
  Center

John W. ("Jack") Rowe
Executive Chair
Health Policy and Management
Columbia University

Donald Royall, M.D.
Chief: Division of Aging and
  Geriatric Psychiatry
Psychiatry, Medicine and
  Pharmacology
University of Texas Health Science
  Center at San Antonio

Khaled Saleh
Associate Professor, Orthopedic
  Surgery and Health Evaluative
  Sciences Division
Division Head and Fellowship
  Director, Adult Reconstruction
University of Virginia

Judith A. Salerno
Deputy Director
National Institute on Aging
National Institutes of Health

Chris Schaffer
Assistant Professor
Department of Biomedical
  Engineering
Cornell University

Nicholas Schork
Director of Research, Scripps
  Genomic Medicines, Scripps
  Health

Teresa Seeman
Professor
Division of Geriatrics/Department of Medicine
University of California, Los Angeles

Jerry Shay
Professor
Cell Biology
University of Texas Southwestern Medical Center

Felipe Sierra
Director
Biology of Aging Program
National Institute on Aging

James Simpkins
Professor and Chair
Pharmacology and Neuroscience
University of North Texas Health Science Center

William Skane
Executive Director
Office of News and Public Information
The National Academies

Richard Sprott
Executive Director
Ellison Medical Foundation

Rob Stephenson
Assistant Professor
Hubert Department of Global Health
Emory University, Rollins School of Public Health

David Stopak
Recruiting Editor
Proceedings of the National Academy of Sciences
The National Academies

Kimberly Suda-Blake
Program Director
Keck *Futures Initiative*
The National Academies

Richard Suzman
Director
Behavioral and Social Research Program
National Institute on Aging
National Institutes of Health

Mercedes Talley
Program Director
W. M. Keck Foundation

Michael Tannebaum
Graduate Science Writing Student
University of Georgia

Charlotte A. Tate
Dean
College of Applied Health Sciences
University of Illinois at Chicago

Heidi A. Tissenbaum
Associate Professor
Program in Gene Function and Expression
University of Massachusetts Medical School

Eric Topol
Director, Translational Science
   Institute
Molecular and Experimental
   Medicine
The Scripps Research Institute

Michael Torry
Director
Biomechanics Research Laboratory
Steadman-Hawkins Research
   Foundation

Iris Tse
Graduate Science Writing Student
Boston University

Ken Turteltaub
Senior Staff Scientist
Chemistry, Materials and Life
   Sciences
Lawrence Livermore National
   Laboratory

Rachel VanCott
Graduate Science Writing Student
Massachusetts Institute of
   Technology

Charles Vest
President
National Academy of Engineering

Helen Vlassara
Mount Sinai Professor in Diabetes
   and Aging
Division of Experimental Diabetes
   and Aging
Mount Sinai School of Medicine

Ken Wachter
Professor and Chair
Demography and Statistics
University of California, Berkeley

Jonathan Wanagat
Senior Fellow/Acting Instructor
Gerontology and Geriatric
   Medicine
University of Washington

David J. Waters
Professor and Associate Director
Center on Aging and the Life
   Course
Purdue University

Molly Webster
Graduate Science Writing Student
New York University

Heinz-Ulrich G. Weier
Staff Scientist
Life Sciences Division
Lawrence Berkeley National
   Laboratory

Mike Weinrich
Director
National Center for Medical
   Rehabilitation Research

Mary White
Chief Epidemiology and Applied
   Research Branch
Division of Cancer Prevention and
   Control
Centers for Disease Control and
   Prevention

Catherine Wolkow
Tenure-track Investigator
Laboratory of Neurosciences
National Institute on Aging-
    Intramural Research Program,
    National Institutes of Health

Savio L-Y Woo
Whiteford Professor and Director
Musculoskeletal Research Center/
    Department of Bioengineering
University of Pittsburgh

Woodring Wright
Professor
Cell Biology
University of Texas Southwestern
    Medical Center

Anatoli Yashin
Professor
Center for Population Health and
    Aging
Duke University

Carl Zimmer
Freelance Writer

Thomas Zimmerman
Research Staff Member
Computer Science
IBM Almaden Research Center